ROUTLEDGE LIBRARY EDITIONS:
POLITICAL SCIENCE

THE LABOUR THEORY OF CULTURE

THE LABOUR THEORY OF CULTURE

A Re-examination of Engels's Theory of Human Origins

By

CHARLES WOOLFSON

Volume 42

Routledge
Taylor & Francis Group
LONDON AND NEW YORK

First published 1982

This edition first published in 2010
by Routledge
2 Park Square, Milton Park, Abingdon, Oxon, OX14 4RN

Simultaneously published in the USA and Canada
by Routledge
270 Madison Avenue, New York, NY 10016

Routledge is an imprint of the Taylor & Francis Group, an informa business

British Library Cataloguing in Publication Data
A catalogue record for this book is available from the British Library

ISBN 10: 0-415-49111-8 (Set)
ISBN 13: 978-0-415-49111-2 (Set)

ISBN 10: 0-415-55583-3 (Volume 42)
ISBN 13: 978-0-415-55583-8 (Volume 42)

The labour theory of culture

a re-examination of Engels's theory of human origins

Charles Woolfson
Department of Social and Economic Research
University of Glasgow

Routledge & Kegan Paul
London, Boston and Henley

First published in 1982
by Routledge & Kegan Paul Ltd
39 Store Street, London WC1E 7DD,
9 Park Street, Boston, Mass. 02108, USA and
Broadway House, Newtown Road,
Henley-on-Thames, Oxon RG9 1EN
Photoset in 10/12 point Baskerville by
Saildean Limited, Surrey
and printed in Great Britain by
Ebenezer Baylis & Son Ltd, Worcester

Library of Congress Cataloging in Publication Data

Woolfson, Charles, 1946-
The labour theory of culture.
Bibliography: p.
Includes index.
1. Communism and anthropology. 2. Human evolution. 3. Labor and laboring classes – History. 4. Engels, Friedrich, 1820-1895 – Anthropology. I. Title.
HX550.A56W66 306 81-19960

ISBN 0-7100-0997-6 AACR2

For John Foster and William Coyle

Contents

Acknowledgments

The following people provided helpful comments and suggestions: David Hamilton, Bridget Fowler, Barbara Littlewood, John Hoffman, Jeffrey Weeks, Henry Mins and Bruce Trigger. Kathleen Clark typed and retyped more times than she probably cares to remember.

Despite the inevitable selectivity in this brief text, the writer sincerely hopes that he has not misrepresented the views of those whose work has been cited.

Introduction

What is surprising, given the concern of Marxist theory with the issue of human development, is the almost total neglect by Western Marxists of recent advances in archaeology, palaeontology, linguistics and such related fields as primatology and anthropology, all of which have raised new and important questions concerning the very origins and nature of the human species.[1] Despite an increasing penetration of Marxist ideas in some quarters, little has been said by Marxists about the emergence of early human beings since Engels attempted to outline the unique path of progressive evolutionary advance towards humankind from our anthropoid ancestors in a brief unfinished essay, written in 1876, entitled 'The Part Played by Labour in the Transition from Ape to Man'.[2] Until very recently, only a handful of scholars have attempted to pursue a materialist interpretation of human development of the kind advanced by Engels.[3] Yet Engels's essay raises an issue central to the whole premises of Marxist theory, namely the formative role of labour in human development and the genesis of human culture.

Nowadays the work of Engels, when not simply ignored, is otherwise subjected to vigorous attack.[4] Presumably this is intended in part to prove yet again that all the blame for the dogmatic and mechanistic 'errors' of Marxism should be laid at the door of Marx's co-worker and not at that of the 'great man' himself. Thus rid of the ballast of Engels, Marxist theory can float even higher in the rarified atmosphere of theoreticism. It is time, however, to come down to earth – and in the most literal sense. In the last two decades exciting evidence concerning the birth of humanity has been dug out of the soil and rock faces of Europe, Africa and Asia. Moreover, in scientific laboratories and in the field, experiments and observations of animal behaviour, including that of our nearest primate relatives, have revealed wholly unexpected patterns of behaviour with profoundly suggestive implications for any theory of how human beings might have developed. So, too, the surviving remnants of simpler forms of society than our own have raised questions about the social nature of human beings and the manner in which they organise their survival. With the whole future evolution of humanity now at stake, the study of our past evolution as a species is more than a fascinating diversion. Our possibilities as well as our limitations are revealed in this study. Yet, with a few notable exceptions, Marxists in particular have remained indifferent to the questions which these researches have raised.[5] A work that spans the whole period of human evolution is entirely beyond the competence of this writer. Indeed, following Engels, only the very first conjectural stage (although perhaps most crucial 'transitional' phase, in as much as all subsequent development was shaped by it) is sketched here. Yet if the role of conjecture still remains embarrassingly large, how much more so this was for Engels.

Much of Engels's essay appears to be simply speculative. The evidence available at the time comprised little more than a few fragments of jaw-bone. In the last quarter of the nineteenth century it was still possible to discern only the most blurred outlines of our prehistoric progenitors. In the last quarter of the twentieth century the role of speculation in some ways looms even larger, although certain issues which were unresolved through sheer lack of evidence have been clarified. It is now not so much the results by themselves of this or that piece of research which are important, but their interpretation. Very often it is the interpretation of existing results which shapes the whole direction which future research will take. What Engels achieved over a century ago was to suggest an entirely new emphasis in the study of human evolution based upon a close study of the existing theories which were reinterpreted within the perspective of dialectical materialism that Marx and Engels both shared as the fruits of their joint labour over many years.[6] In that sense Engels provided the guidelines for a future Marxist interpretation of human evolution. In their own day both Marx and Engels eagerly seized upon the work of Charles Darwin, as representing the most advanced scientific understanding of the process of evolutionary development.[7] When Darwin eventually turned his attention to humankind in *The Descent of Man,* over a decade after the publication of the *Origins of the Species,* he provided Engels with a starting-point for a dialectical reworking of Darwin's ideas that generated a number of formulations and hypotheses which still have relevance for contemporary scholarship.[8]

Darwin, like many of the writers whose work is examined below, was more anxious to stress the *continuities* between anthropoids and man. Engels, on the other hand, considered also the qualitative *discontinuities* introduced by the emergence of social labour, what it was that was *special* about the development of human beings as a species. While it is true that certain details of his work have been refuted, had Engels been alive today he would undoubtedly have greeted the latest scientific data with alacrity. It is the intention of this book to show that the broad outlines of Engels's theory are, by and large, confirmed by contemporary research, and that in this respect Engels's essay is a brilliant scientific anticipation of what is now thought, by some writers at least, to be the likely pattern of early human evolution. It will be suggested, moreover, that if we are to succeed in providing the kind of interpretation of these researches which will enable their broader significance to be grasped, it is necessary to reappropriate Engels's theoretical legacy in the kind of dialectical reconstruction of recent discoveries that Engels himself proposed.

First of all, therefore, in chapter 1 the main ideas contained in Engels's essay are further elaborated. In chapter 2, some of the archaeological data describing the tools and fossil remains of the earliest hominids are commented upon and certain suggestions are made as to how the first humans might have begun to develop physically as creatures that gradually adapted to labour and, in turn, were themselves adapted by labour. The forms of rudimentary social organisation within which members of the species possibly learned the advantages of co-operation are discussed in chapter 3. It is argued that this was the most likely precondition for subsequent cultural advance, rather than the competitiveness and aggression which some palaeontologists have attempted to read into the prehistory of humanity. In chapter 4, various field studies of higher primates are reviewed, particularly with regard to behaviour involving the use and manufacture of tools. Some writers have argued on this basis that such behaviour is equivalent to labour activity among humans. In chapter 5, more evidence is presented which would seem to blur still further the line of demarcation between human and animal behaviour, specifically in the laboratory investigations of linguistic skills of primates. However, by a careful examination of the specifics of activity involved in human tool-behaviour and linguistic behaviour it can be shown that, taken together, these would have *decisively altered* the way in which early humans interacted with each other and their environment. Labour as such cannot be performed by animals other than humans and this is the real basis of their uniqueness. In other words, as is argued in chapter 6, fairly early on in human development, perhaps much earlier than many may have imagined hitherto, members of our species had begun to plan and organise their lives precisely because they had developed some form of spoken language in the labour process. Language also provides initial access to the dimension of time and, therefore, it is suggested, the historical passing on of a cultural inheritance of acquired skills and knowledge won through labour becomes possible. The final chapter takes forward the question of the interconnection of tool-making and language, and examines, within the framework of a Marxist psychology, the crucial connection between hand and brain which is forged through labour and opens up to human beings alone the possibility of mastering both themselves and their world.

1 Engels and human origins

The starting-point for a materialist analysis of human development is the manner in which mankind produces its means of subsistence. What is produced and how it is produced identifies what Marx had called 'the species-character' of human beings.[1] For Engels, as for Marx, the central feature distinguishing human beings from other species is *labour.* In *Capital* Marx had defined labour in terms of the contradiction between the subject, man, and the object, nature.

> Labour is, in the first place, a process in which both man and Nature participate, and in which man of his own accord starts, regulates and controls the material reactions between himself and Nature. He opposes himself to Nature as one of her own forces, setting in motion arms and legs, head and hands, the natural forces of his body, in order to appropriate Nature's production in a form adapted to his own wants. By thus acting on the external world and changing it, he at the same time changes his own nature. He develops his slumbering powers and compels them to act in obedience to his sway.[2]

Marx was here describing labour in its developed form. What Engels attempted to do was to show how humanity developed up to that point where labour became the typical and enduring form of human activity and the main impetus to its further development. Engels, indeed, like Marx regarded human labour as central to mankind's overall development as a species. 'It is the primary basic condition for all human existence, and this to such an extent that, in a sense, we have to say that labour created man himself.'[3]

Engels's position may be briefly summarised as follows: with the adoption of an erect posture and bipedal locomotion, the hand of early man was *freed* to acquire an increasing facility in tool-use and tool-making which led in turn, over time, to further changes in the structure of the hand, such that the hand became not only 'the organ of labour', but also *'the product of labour'*.[4] Elsewhere in *Dialectics of Nature* Engels contrasts the development of the human hand through labour with the 'tools' and 'production' of animals.

> The specialisation of the hand – this implies the *tool*, and the tool implies specifically human activity, the transforming reaction of man on nature, production. Animals in the narrower sense also have tools, but only as limbs of their bodies: the ant, the bee, the beaver; animals also produce, but their productive effect on surrounding nature in relation to the latter amounts to nothing at all. Man alone has succeeded in impressing his stamp on nature.[5]

Thus for animals 'tools' are, in the main, biologically pre-given and consolidated in the natural development of each member of the species as part of their physical make-up. Their interaction with nature in the form of direct responses to environmental stimuli is in this sense a generally passive and 'one-sided' affair.[6] Mankind's relation to nature, on the other hand, is an active transformative one in which extra-bodily, artificially produced and indirect means of adapting nature to its requirements are frequently employed. Marx also was clear that, as an aspect of the labour process, the making of tools, substituting so to speak for the lack of natural equipment, provided a key to the unique mode of interaction between humans and their environment. Again, in *Capital* Marx writes,

> An instrument of labour is a thing, or a complex of things which the labourer interposes, between himself and the subject of his labour, and which serves as the conductor of his activity ... As the earth is his original larder, so too it is his original tool house. It supplies him, for instance, with stones for throwing, grinding, pressing, cutting &c ... No sooner does labour undergo the least development, than it requires specially prepared instruments ... For use and fabrication of instruments of labour, although existing in germ among certain species of animals, is specifically characteristic of the human labour-process, and Franklin defines man as a tool making animal.[7]

The original tools of early humans were, as is suggested in the following chapter, fairly primitive efforts. Nevertheless, they were profoundly significant in terms of what they represented for the future development of human capacities.

> The mastery over nature, which begins with the development of the hand, within labour, widened man's horizon at every new advance. He was continually discovering new, hitherto unknown, properties in natural objects. On the other hand, the development of labour necessarily helped to bring the members of society closer together by multiplying cases of mutual support, joint activity, and by making clear the advantage of this joint activity to each individual. In short, men in the making arrived at the point where *they had something to say* to one another.[8]

Speech provided the necessary symbolic apparatus with which to begin to organise, preserve and transmit the collective labour experience of humanity. 'First comes labour, after it and then side by side with it, articulate speech – these were the two most essential stimuli under the influence of which the brain of the ape gradually changed into that of man.'[9] Finally, the development of the brain in turn further stimulated both tool-related activity and language, which became a more precise instrument for planning future labour activity.

> The reaction on labour and speech of the development of the brain and its attendant senses, of the increasing clarity of consciousness, power of abstraction and of judgement, gave an ever renewed impulse to further development of both labour and speech.[10]

Here Engels was positing a relationship of positive feedback between the general development of mental faculties and the continuous increase in the efficiency and the quality of human labour. As the planning of future activity, the identification of the aims of labour, the assessment of properties of objects, and the division of tasks within the labour process, slowly came into being, they did so within an increasingly social and co-operative context. Earliest man became *humanised* through labour. The feedback was not merely positive but cumulative, with labour activity providing the starting-point for general human advance.

> By the co-operation of hands, organs of speech and brain, not only in each individual but also in society, human beings became

> capable of executing more and more complicated operations, and of setting themselves, and achieving higher and higher aims. With each generation labour itself became different, more perfect, more diversified.[11]

In principle at least, the productive techniques of early humans could be refined and augmented. Not so among animals who have no way other than through the adaptation of their own bodies of adding to the efficiency with which they produce. Human beings can create their own special access through speech and tool-making to an ever widening pool of social information about the problems met with and overcome in securing their means of subsistence. Tool-making and speech, therefore, could be said to provide the twin foundations for the whole subsequent development of human culture. Appreciation of environment, knowledge of the seasons, observed regularity in the habits of prey, an understanding and selection of materials and the processes of manufacture of tools to serve different purposes, all that became the field for the successive expansion of increasingly *human* action as mankind sought to bring both the external world of nature and its own activity under conscious control.

> In short, the animal merely *uses* external nature and brings about changes in it simply by his presence; man by his changes makes it serve his ends, *masters* it. This is the final, essential distinction between man and other animals, and once again it is labour that brings about this distinction.[12]

The general processes of such mastery may be designated by the term culture. Cultural development then, is expressed through the accumulated material achievements of social practice and the social and spiritual achievements conditioned by them. At every stage of historical development, culture is a measure of man's humanisation, the degree to which he has separated himself from his animal origins, the extent to which he has humanised nature and his own being as a part of nature through his labour activity. It can be said that what Engels outlined was in effect a 'labour theory of culture', culture being as unique to mankind in terms of the possibilities for development which its possession offers, as is labour itself.

For Marx and Engels these possibilities could only be realised fully in a future form of society in which there would be the conscious organisation of social production in a planned way. Only then, as labour was freed from the distorting effects of exploitative modes of production, would the results of human activity correspond with their intentions.[13]

Darwin's understanding of human mastery of the environment was a far more limited one than this. Lacking a clear conception of labour he also lacked the necessary grounds for identifying precisely the qualitatively new characteristics that placed human evolutionary development on a new path. Darwin saw only the quantitative and gradualist aspect of this question, claiming that there was no fundamental difference between man and higher mammals in their mental faculties, but rather the difference between them consisted

> solely of his almost infinitely larger power of associating together the most diversified sounds and ideas ... the difference in mind between man and the higher mammals, great as it is, certainly is one of a degree and not of kind.[14]

This essentially idealist proposition still commands considerable allegiance today, although the arguments in its favour are presented in a more sophisticated way. It usually consists in marking out a continuum of behaviours and abilities. By stressing continuities at the expense of discontinuities, human activity is seen as merely a quantitative extension of that of other species. The qualitative uniqueness of mankind is contested on every side and the role of labour in defining that uniqueness is submerged.[15]

It is not possible here to trace out the full lines of Marx and Engels's thinking on the question of the eventual outcome of the historical process of human development. The concern of this work is rather to examine what new evidence there is which might throw some light on the very first basic step in the evolution of humanity as represented by the inauguration of labour activity. Darwin performed an important service by interesting us in, as Marx put it, 'the history of Nature's Technology', that is of the formation of 'productive organs' of plants and animals which 'serve as instruments of production for sustaining life'.[16] Marx then adds, by analogy

> Does not the history of the productive organs of man, of organs that are the material basis of all social organisation, deserve equal attention? And would not such a history be easier to compile,

> since as Vico says, human history differs from natural history in this, that we have made the former, but not the latter? Technology discloses man's mode of dealing with Nature, the process of production by which he sustains his life, and thereby also lays bare the mode of formation of his social relations, and of the mental conceptions that flow from them.[17]

There are those who perhaps with good reason would dispute whether, given the relatively incomplete state of scientific knowledge, it is yet very much 'easier' to compile the history of early humanity's 'productive organs'. For the greater part of human history spanning a period of literally millions of years the evidence of the technology of mankind which has been uncovered to date is still infinitesimally small. What we know of early human social relations and of their 'mental conceptions' is even less. Nevertheless, the fossil record does at least provide us with certain clues as to what might have been. Whether the erection of a whole theory of human origins of the kind Engels proposed is justified, on such a limited data base, readers must judge for themselves, since, even without an explicitly Marxist interpretation, the fossil record is now the subject of acute scientific controversy.

2 The fossil record

The earliest hominid fossils, which date as far back as 14 million years before the present, show no association with tools at any of the forty odd sites discovered since G. Edward Lewis first uncovered a jaw-bone of *Ramapithecus* in the Siwalik Hills in north-west India in 1932.[1] This is either because the first hominids did not employ tools or, as seems more likely, given what we know of subsequent hominid evolution, because such tools were made of wood and other perishable materials which have not been preserved in the fossil record. However, although no tools have been found the earliest hominid does differ in certain crucial aspects from his ape-like cousins, particularly in his dentition and the shape of his jaw. Commenting on this, Simons remarks:

> The proportions of the jaw indicate a foreshortened face. The size ratio between front teeth and cheek teeth is about the same as it is in man. (The front teeth of living apes are relatively large.) Estimating from the size of its socket the canine tooth was not much larger than the first premolar – another hominid characteristic, opposed to the enlarged canine of the pongids. The arc formed by the teeth is curved as in man, rather than being parabolic, or U-shaped as in the apes.[2]

The gradual reduction of canines to the level of incisors suggests strongly that the teeth of *Ramapithecus* were no longer employed for purposes of killing and tearing. Rather, 'extra-somatic' tools had begun to be substituted for these functions. As Washburn has pointed out, 'Small canines and incisors are biological symbols of a changed way of life; their primitive functions are replaced by hand and tool'.[3] Simons suggests that these changes could well be the result of a dietary shift consequent upon foraging on the ground at the forest edge for roots and nuts in a wooded – savanna environment, after abandoning an arboreal way of life.

> Even as they were still living in forests and on the margins of forests they had given up feeding in the trees in favour of feeding on the ground. Indeed, *Ramapithecus* was already at home away from the treetops in the open woodland and savanna.[4]

Napier suggests that *Ramapithecus* probably occupied an environment intermediate between forest and open grassland, a kind of half-way house niche which is today occupied by many primates. 'It was probably in this transitional environment that man's ancestors learned to walk on two legs. In all likelihood, however, they only learned to stride when they later moved into the open savanna.'[5]

What brought about the desertion of a predominantly arboreal way of life is largely a matter of speculation. Hockett and Ascher in their essay 'The Human Revolution' cite the essentially 'conservative' but not entirely implausible Romer's law as underlying the shift, or rather 'eviction', of man's ancestors from the tree-dwelling environment.[6] Briefly, Romer's law argues that evolutionary adaptations may occur primarily as a result of an attempt to *conserve* a pre-existing mode of life in changing environmental conditions. The change which Hockett and Ascher see as crucial is climatic, whereby vegetation thinned out and previously continuous forest was converted into open savanna and scattered clumps of trees. Some bands of our pre-human ancestors were perhaps caught in isolated groves of diminishing extent. Those that survived were those who could travel across open ground, not in order to begin a new way of life out of the trees, but to make their way to where traditional arboreal life could be continued. Perhaps also population pressures within a diminishing grove might force out less powerful bands and cause migrating bands to move to new groves where subsequent competition might ensue with the existing population.[7]

> Thus, in the long run the trees would be held by the more powerful, while the less powerful would repeatedly have to get along as best they could in the fringes of the forest or in the open country. Here is a double selective process. The trees went to the more powerful, provided only that they maintained a minimum ability to traverse open country when necessary: some of the successful ones were ancestral to the great apes of today. Our own

> ancestors were the failures. We did not abandon the trees because we wanted to but because we were pushed out.[8]

However this may be, these creatures may well have succeeded in effecting quite novel forms of adaptation to their environment with the aid of tools. It is suggested by Coursey that, in coping with the exigencies of their new environment, tool-usage would probably have become crucial to the survival of early humans.[9] Coursey argues, taking issue with Jolly, that if *Ramapithecus* subsisted on a diet mainly of grass seeds this would not provide sufficient stimulus for tool-use.[10] Even the crudest form of digging stick, however, would greatly increase the efficiency with which water-bearing plant foods could be extracted, particularly where they grew deeply beneath the ground or had spiny stems and roots as mechanical defence mechanisms. Such behaviour would be in advance of anything achieved by the anthropoids and could well have assumed immense evolutionary significance. Coursey notes,

> It is suggested therefore that it is in association specifically with plants of this type that the increasingly regular, frequent and ultimately obligate use of the digging stick originated, as the ancestral hominids progressively occupied the savanna habitat and adopted a terrestrial habitat. The ancestral pongids, remaining arboreal within the forest, did not have the same incentive to become obligate tool-users and so divergence occurred.[11]

It is not too difficult to imagine that where these tubers and roots became scarce or perhaps dried out under conditions of prolonged drought, sticks that might at first have been employed as digging tools may then have been used to kill smaller game thus setting in motion a new trend towards hunting in order to supplement diet. The question of hunting is taken up again later. For the moment, lacking preserved fossils of early wooden tools the foregoing is mainly suppositional. However, it can be argued that while the divergence between humans and anthropoids may well have had its initial foundation in the preceding forms of tool-use, the decisive 'moment' of their differentiation appeared with tool-*making*. It is not impossible that *Ramapithecus* had already taken at least the first steps towards extending its activities in this direction, particularly if such creatures had begun to adopt bipedal locomotion and an omnivorous diet. Brace and Montagu comment in this regard:

> An organism with freed hands can not only manufacture tools, but also carry tools and other objects with it as a substitute for anatomical inadequacy. The regular manufacture of tools and the reliance on them as a primary means of survival immediately implies the existence of a complexity of learned behaviour and traditions, which we recognise under the name of culture. Even without any further evidence then, anatomical indications for bipedalism should lead us to suspect that culture was the primary means of adaptation, and since we define 'human' as the organism whose primary means of adaptation is culture, then our biped must by definition be human, however primitive.[12]

Although Lewis had suggested as far back as 1934 that *Ramapithecus* might well belong to the *Hominidae* it was not until 1961 that the jaws of a similar hominid were uncovered by Louis Leakey at Fort Ternan in south-western Kenya.[13] Commenting on Leakey's find, Simons remarked in 1964,

> The conclusion is now almost inescapable, in late Miocene to early Pliocene times both in Africa and India an advanced hominoid species was differentiating from more conservative pongid stock and developing important hominid characteristics in the process.[14]

Further finds in the 1970s of *Ramapithecus* fossils in Greece, Turkey, Hungary and Pakistan suggest that *Ramapithecus* was truly cosmopolitan, a conclusion which Louis Leakey's son and successor, Richard, seems to accept in a recent popular review of these finds.[15] Unfortunately, we have to wait until much nearer the present before we find more convincing evidence of either bipedalism or the inception of a human cultural tradition worthy of its name.

Between *Ramapithecus* and the appearance of the next nearest human ancestor, *Australopithecus,* there is a gap in the fossil record of some 4 million years, aptly referred to by Leakey and Lewin as 'the fossil void'.[16] Thanks to the remarkable series of discoveries by Louis Leakey and his wife Mary in the lower strata of the Olduvai Gorge in Tanzania, we now have the first conclusive evidence of tool-making early man. In 1959 the Leakeys found fragments of the skull of an early small-brained man-ape in rock which was laid down about 1.75 million years ago.[17] What was remarkable about this specimen – named *'Zinjanthropus'* by Leakey – was that it was found in clear association with stone tools including a hammerstone and waste flakes, indicating a manufacture process, the first evidence of tool-making by early humans.[18] The Olduvai deposit not only contained simple tools but also the bones of rats, mice, frogs and a juvenile pig and antelope suggesting the hunting of small animals. Since the Olduvai Bed I discoveries, important finds have been made both at Olduvai and at further locations in East Africa along the line of the Great Rift Valley including Lake Turkana in Tanzania, Koobi Fora in Kenya, the Omo Valley and Hadar in Ethiopia.[19]

The work of Binford and Binford suggests that the different types of activity which might have comprised the behaviour of early hominids is likely to account for the variations in tool assemblages and other remains found at these sites.[20]

> In technologically simple societies we can distinguish two broad classes of activities: extraction and maintenance. Extraction involves the direct procurement of food, fuels and raw materials for tools. Maintenance activities consist in the preparation and distribution of foods and fuels already on hand and in the processing of raw materials into tools. Since the distribution of resources in the environment is not necessarily related to the distribution of sites providing adequate living space and safety, we would not expect extraction and maintenance activities to be conducted in the same places.
>
> Base camps are chosen primarily for living space, protection from the elements and central location with respect to resources. We would expect the archaeological assemblages of base camps to reflect maintenance activities: the preparation and consumption of food and the manufacture of tools for use in other less permanent sites.
>
> Another settlement type would be a work camp, a site occupied while smaller social units were carrying out extractive tasks. Archaeologically these would appear as kill sites, collecting stations and quarries for extracting flint to be used in toolmaking. The archaeological assemblages from these sites should be dominated by the tools used in the specific extractive tasks. If a work camp were occupied for a rather long period and by a fairly large subgroup, we would anticipate that some maintenance activities would also be reflected in the archaeological remains.[21]

An example of what appears to be a base camp at Olduvai is referred to in the following chapter. There are also, however, several well-documented examples of work camps in the east African finds including kill sites and a 'tool-factory'.

From excavations at Koobi Fora by Isaac and others, stone tools have been discovered at what appears to be a butchering site of a hippopotamus, the so-called hippopotamus/artifact site (HAS).[22] While Isaac is cautious about whether it was an australopithecine or some other species of bipedal hominid that made the tools discovered at the HAS site, it seems clear from the tool assemblage that early hominids carried with them suitable raw materials for the manufacture of stone tools on site for distances of perhaps 3 or 4km. Isaac notes:

> The sediments where we found these artifacts contain no stones larger than a pea. Thus it seems clear that the makers of the tools had carried the stones here from somewhere else. The association between the patch of artifacts and the hippopotamus bones further suggests that toolmakers came to the site carrying stones and hammered off the small sharp-edged flakes on the spot in order to cut meat from the hippopotamus carcass.[23]

This would indicate that particular materials were chosen in advance with their working properties in mind and that some systematic knowledge of the most efficient means of performing labour might have begun to be accumulated. The HAS site suggests the general character of a labour process involving the pre-selection of materials for the manufacture of instruments of labour, specifically hammerstones and core tools. In their application to the tasks of cutting and scraping skin, sinew, flesh and bone materials the small sharp raw flakes produced would be very effective.

From Olduvai Bed II, in deposits about 1.6 million years old, comes evidence of a further type of work-camp site, the MNK chert 'factory' site to which stone materials were also imported, in this case for initial preparation as tools.[24] These in turn seem to have been taken to living sites where they were utilised or transformed into shaped tools. Besides a very large number of chert specimens, two hammerstones and working anvil stones were identified.[25] It is suggested that the chert stone materials were brought to the site from sources of no more than 1km. distance and that the absence of bones or other remains usually associated with living sites suggests its specific function as a factory site.[26] This would imply the kind of spatial ordering of the environment by early hominids in the manner described by the Binfords.

While it is certainly true that the early Oldowan tools were of fairly simple manufacture and design and remained so for tens and indeed hundreds of thousands of years, it has been suggested that the mere outward appearance and design of the tools may conceal the fact that important cultural advances were being made. Jones has shown that in assessing lower Palaeolithic bifaces from Olduvai Beds III and IV, that is up until about 0.7 million years ago, the amount of retouching and secondary flaking in tools is a poor guide to the quite complex character of the labour that produced them.[27] Jones manufactured experimental bifaces by means of freehand percussion employing basalt hammerstones on stone-tool blanks obtained from the same sources as those used by the inhabitants of Olduvai. Different raw materials were shown to have different flaking and edge qualities. Thus the optimum edge of a cutting tool might not necessarily be obtained by secondary flaking, while some materials were more easily blunted in use than others, or required more time and effort to manufacture or retouch. Increased mastery of the raw materials for tool-making, therefore, does not necessarily result in stone tools that are more refined in their outward appearance. Rather, the qualities of the materials and the likely uses to which they were put must be taken into account. Despite the apparent lack of trimming on stone tools early humans may well have been slowly and steadily accumulating valuable information about how to cope with their environment and utilise its resources in the most economic and labour-saving way.

Similarly, while it is not necessarily possible to infer developmental stages directly from the tools, caution should also be exercised in inferring distinctive cultural traditions from variations in the range of tool types found at sites of roughly comparable age. These may either reflect simple differences in the availability of raw materials or differences in the kind of activities conducted at various sites.

Jones has since performed experimental butchering of such animals as goats and zebras with the aid of stone tools in an effort to test out the working properties of different tools more fully.[28] Interesting though such reconstructions of possible labour activity are, perhaps the most promising development for a precise analysis of the manufacture and use of stone tools is the employment of electron microscopes to analyse patterns of surface wear on tools.[29] Unfortunately, to date these studies have concentrated on tools of much more recent origin. However, such advanced techniques as these, together with radio-carbon dating techniques, should eventually enable a much clearer picture to be drawn of the pace and form of the evolution of tool-making and labour by early humans.

In South Africa the pioneering work of Dart and his colleagues has continued to yield a whole series of australopithecine remains, beginning with the famous Taung skull in 1924, as well as numerous 'tools'. These are often of uncertain provenance as will be suggested in the following chapter. However, the limestone cave deposits which have yielded these finds do not permit the application of the most advanced scientific dating techniques. The earliest stone tools that have been uncovered so far, from Hadar in Ethiopa by Johanson and his co-workers, predate those of Olduvai and suggest that stone tools were in habitual use by early humans as long ago as 2.5 million years ago.[30] Since it is highly improbable that these were the first stone tools, far less the first tools ever made by humans, it suggests that the origin of human tool-making may well extend as far back as 3 million years and, if *Ramapithecus* was also a tool-maker, possibly much earlier still. As yet, however, more evidence is required before this question can be answered with certainty.

The further discovery at Lake Turkana of a large-brained hominid fossil remains by R. Leakey in 1972 – the much publicised KNM-ER 1470 – was also initially dated at about 2.5 million years.[31] This skull, claimed Leakey, placed KNM-ER 1470 in the genus *Homo* having a cranial capacity of 775cm^3 – substantially bigger than that of australopithecines which ranges from 422 to 539cm^3. This suggested a marked growth in cranial capacity much earlier even than implied by Louis Leakey's previous discoveries, and forced back the division between *Homo* and *Australopithecus* by at least a million years. There were, moreover, stone tools found in association with the Turkana fossils, further circumstantial evidence that *Homo* was the tool-maker. Subsequently doubt has arisen as to the precise age of the KBS tuff which provides the reference point for dating these remains.[32] It now seems more likely that both the skull and the stone tools are no older than similar finds at Olduvai. This means that it is still not possible to argue, as Leakey has implied, that tool-making was preceded by or conditional upon a *prior* expansion of cranial capacity. Whether, however, there is a single line of human evolution leading from *Ramapithecus,* through *Australopithecus* to *Homo,* or, as Leakey has tried to argue, there was an early evolutionary divergence between *Homo* and *Australopithecus* with the latter becoming extinct offshoots of the evolutionary tree, is currently a matter of competing interpretations, although the writer favours a broadly unilineal approach.

In this respect the remains of several dozen hominids from Hadar in Ethiopia dated over 3.5 million years ago are of some relevance. These were originally held to suggest the co-existence of two types of australopithecine 'robust' and 'gracile' as well as *Homo* perhaps as recently as 3 million years ago, but now seem to be variations of the same australopithecine population.[33] Particularly interesting is the recovery of 40 per cent of a skeleton of a female australopithecine, the now famous 'Lucy', including fossil hand-bones which suggest that these hominids may have rivalled modern human powers of dexterity and fragments of lower-limb bones, including pelvic and thigh bones, which suggest full bipedalism.[34] Unfortunately, no stone tools have been found in direct association with these fossils although they may well have used other less durable implements. If the hands were as dextrous as the fossil evidence implies this reinforces the suggestion that some form of tools were manufactured by developing hominids from an early date.

Perhaps the most dramatic discovery of all is the 1979 report of two, possibly three, sets of Pliocene hominid footprints fossilised in a river bed at Laetoli in northern Tanzania, near Olduvai. These have been dated at 3.5 – 3.75 million years ago and, in the words of Leakey and Hay, they reveal

> that the Pliocene hominids at Laetoli had achieved a fully upright, bipedal and free striding gait, a major event in the evolution of man which freed the hands for tool-making and eventually led to more sophisticated human activities. Moreover, evidence supplied by cranial parts of the somewhat later but related hominid fossils from the Afar in Ethiopia (dated between 2.6 and 3 million years) indicates that bipedalism outstripped enlargement of the brain. To have resolved this issue is an important step in the study of human evolution, as it has long been the subject of speculation and debate.[35]

What Leakey and Hay's evidence suggests is that the feet which left these prints belonged to hominids whose walk was very close to that of modern humans, thus suggesting that while doubt remains concerning *Ramapithecus,* probably *Australopithecus* was fully bipedal at a very early date. The authors comment that, 'the longitudinal arch of the foot is well developed and resembles that of modern man, and the great toe is parallel to the other toes.'[36] It seems fair to assume that for the foot to have evolved changes must have already also occurred in the pelvic structures; indeed, such evidence is reviewed by Napier based on later australopithecine remains.[37] More recently, on the basis of *biomechanical* analysis, Lovejoy and his colleagues have suggested that the free-striding gait of *Australopithecus* was equal, if not superior, to its human successors among whom the modification of the pelvis to accommodate the birth of encephalised infants may actually have reduced the efficiency of their stride.[38]

Napier has also commented that bone fragments from the Oldowan hand show that, although a precision grip in its fullest development might not be wholly within its capabilities, *Australopithecus* would probably have little difficulty in making the tools found with it by means of a power grip. Such tools were, however,

> little more than pebbles modified in the simplest way by striking off one or more flakes to produce a chopping edge. This technology could not have required either a particularly large brain or a hand of modern proportions.[39]

Day, while taking issue with aspects of Napier's reconstruction, comments cautiously in a review of this material, 'the hand was powerfully built, capable of strong finger flexion and possessed of an opposable thumb.'[40]

The prehensibility of the specifically human hand, the ability to conduct finer manipulation, is conditional upon the full development of a precision grip itself requiring a fully opposable thumb, whereas the simple power grip does not require a hand of essentially human form and proportions. The evidence would seem to indicate, therefore, that the earliest tool-making was conducted by a hand that had still to undergo considerable modification in a human direction and that in certain crucial respects, like the brain itself, only became fully developed long *after* the initial inception of tool-making.[41] Napier comments:

> The inception of toolmaking has hitherto been regarded as the

> milestone that marked the emergence of the genus *Homo.* It has been assumed that this development was a sudden event, happening as it were almost overnight, and that its appearance was coincidental with the structural evolution of a hominid of essentially modern human form and proportions. It is now becoming clear that this important cultural phase in evolution had its inception at a much earlier stage in the biological evolution of man, that it existed for a much longer period of time and that it was set in motion by a much less advanced hominid and a much less specialised hand than has previously been believed.[42]

Although controversy remains over the precise dating for the first likely emergence of *Homo,* the evidence suggests that stone-tool making had become a habitual part of a developing human way of life *prior to* the major cranial expansion and other anatomical developments which have often been taken to mark the emergence of man.[43] The advantages that the application of tools to their environment would have given their users were seemingly sufficient for them to begin to organise their activities around these. Nevertheless, as Napier suggests, it was not simply a question of biological evolution suddenly being supplanted by cultural evolution. Rather, the development of each was shaped by the other for many millions of years until, fairly recently in evolutionary terms, humanity assumed a predominantly cultural mode of adaptation. Whatever may have been its subsequent pace of development, whether or not the initial emergence of such a novel and significant form of behaviour as tool-making was quite as gradual as Napier seems to imply, is a question which will be returned to.

This chapter has suggested, however, that from fairly early on in human evolution the leading edge of its development was provided by the first tentative cultural innovations embodied in tools. The adoption of tools began to shape not only its environment but also, as it were, the very structure and functioning of mankind's physical organism. As Geertz has put it:

> The fact ... that cultural development was underway well before organic development closed, is of fundamental significance for our view of the nature of man. He becomes, now, not just the producer of culture, but in a specific biological sense, its product.[44]

The implications for Washburn, who has also long held this view, are clear, especially if, as seems justified, his comments on tool-use are taken to imply the refinements of tool-making as well.

> The fossil record thus substantiates the suggestion first made by Charles Darwin, that tool use is both the cause and the effect of bipedal locomotion. Some very limited bipedalism left the hands sufficiently free from locomotion functions so that stones or sticks could be carried, played with and used. The advantage that these objects gave to their users led both to more bipedalism and to more efficient tool use ... Selection is based on successful behaviour, and in the man-apes the beginnings of the human way of life depended on the learned skills of tool-using. The success of the new way of life based on the use of tools changed the selection pressures on many parts of the body, notably the teeth, hands and brain, as well as on the pelvis. But it must be remembered that selection was for the whole way of life.[45]

In the following chapter the likely forms of social organisation within which early humans developed their 'way of life' are examined.

3 Hunting and gathering

The 'way of life' of evolving hominids can be characterised by the term 'hunting and gathering'. In its developed form it bases itself on the employment of tools in the hunting and butchering of animals and in the gathering and preparation of seeds, roots, berries and other plant foods for sharing and collective consumption by members of the group. What perhaps neither Darwin nor Engels realised was how painfully slowly such a way of life became differentiated from that of the great apes.[1] As to how it took shape, only the broadest inferences can be drawn from the fossil remains of early humans. Nevertheless, certain hypotheses seem at least plausible.

It seems likely, for example, that early humans, unlike their nearest anthropoid cousins, did not feed as they ranged. Rather, a proportion at least of food was collected for consumption at some later time. The transportation of stones over some distance to manufacture tools at butchering sites, for example, would seem to suggest that the product of labour was equally likely to be carried away. Similarly, fruits and plants gathered over a wide area would not be individually consumed on the spot but brought back to the group and shared out among its members. Washburn and De Vore comment:

> It is hard to believe that animals that carried stones did not also carry food and thus initiate an important stage in freeing man from the necessity of eating his food where he found it. Carrying food, whether meat or vegetables, opens the way to storage, sharing, and division of labour, so that the presence of carried stones in australopithecine deposits suggests that their economic way of life may have advanced far in the human direction.[2]

Isaac also notes that the development of this new pattern of activity probably had important social implications.

> Food sharing and the kinds of behaviour associated with it probably played an important part in the development of systems of reciprocal social obligations that characterise all human societies we know about.[3]

What is being described here is a form of exchange and orderly sharing that is unknown in the animal world. Social co-operation rather than dominance must have increasingly determined the character of the distributive mechanism in early human society.

The freeing of the hands for carrying presupposes some form of bipedal locomotion. Whereas for the newborn of monkeys and apes the great toe is important in holding onto the mother, this is excluded by bipedal locomotion which fundamentally alters the shape of the foot.[4] Signs of early evolution in this direction are already clear in the recovered australopithecine remains previously discussed. Upright walking necessitated a narrowed pelvis in order to hold the guts in position and brought changes in the shape of the birth canal. During the Pleistocene, however, the relative size of the brain grew by as much as three times. Selection favoured larger and larger heads. Thus, to compensate for the consequent physical problems of birth, humans gradually became born at an earlier stage of development than apes. The result was an increasing period of dependency and the growth of a more prolonged mother–child relationship.[5]

It has been suggested that australopithecines experienced a childhood dependency period of about as long as modern humans. On the basis of South African finds Mann observes that, 'dental eruption in these fossil hominids followed essentially a modern human pattern, i.e., the teeth developed and erupted in the delayed fashion characteristic of modern children.'[6] According to Mann, such a prolonged period of dependency can only be explained as, 'reflecting the time required for the learning and perfecting of skills necessary in a tool-dependent species.'[7] Such a long socialisation is necessitated also in order that the transfer of socially useful knowledge between one generation and the next may take place. Prolonged childhood and the extension of the mother–child relationship would therefore seem to indicate the growth of human culture.

In terms of the practicalities of everyday living the relative helplessness of the human infant implies some sort of secure 'home base' for its upbringing. So, too, the preparation of tools for future use and the need for some form of protection from the elements would imply the selection and habitation of suitable locations of more than brief duration fairly early on in human evolution. It also seems probable that early hominids ranged far more widely in securing their means of subsistence than did any non-human primates. A 'home base' would provide a site for the exchange of the products and information related to hunting and gathering. Evidence of such early living sites, although not of the duration of their occupation or of the likely number of residents, has now accumulated from several parts of Africa. Here again the Olduvai discoveries are among the most important. Mary Leakey observes with regard to the 'Zinj' living-floor debris at the FLK level:

> It has been suggested in discussion that the central living area may have been enclosed, at least to the south and east, by a thorn fence or windbreak, which would correspond with the barren zone, while objects found on the outside might have been thrown out over the fence by the occupants of the camp.[8]

A further living area is suggested at the DK site.

> It is probable that the stone circle at DK formed the base of a rough windbreak or simple shelter. The two factors that are most suggestive of an artificial structure are the small heaps of piled-up stones that form part of the circle and the fact that occupation debris did not occur in comparable density within the circle and in the surrounding area.[9]

To some extent the interpretation of the assemblages of tools and other remains at sites such as these described above can be assisted by our knowledge of surviving hunter–gatherer societies, although several writers warn of the need for caution when using ethnographic data to elucidate the Pleistocene way of life.[10] Even the most backward of hunter – gatherer groups is much more advanced than anything likely to have been achieved by australopithecines in terms of their physical and intellectual capabilities.

With this caveat in mind there is nevertheless now an extensive literature on hunting-and-gathering societies which can perhaps assist in revealing the kind of general principles which might have underlaid early human society. As Lee points out in a recent monograph on the !Kung San people of the Kalahari,

> Foraging was the way of life that prevailed during an important period of human history. The modern foragers do offer clues as to the nature of this way of life, and by understanding the adaptations of the past we can better understand the present and the basic human material that produced them both.[11]

It has already been suggested that the chief organising principle of the very first human societies can be identified in the regular employment and manufacture of tools in the processes of socially co-operative production. Lee offers a fascinating insight into how, with the aid of a relatively simple technology, one group of hunters and gatherers has succeeded in carrying out the basic tasks of subsistence. Among the !Kung, tools for obtaining water, for gathering and carrying, for hunting and for food-processing identify the major areas of productive activity.[12] The efficiency with which such a range of tools can be utilised, while adequate to their purpose, is ultimately circumscribed by the fact that their degree of mastery over nature is still relatively incomplete. Hunter–gatherers must live with nature more or less 'as a given' and adapt their social organisation to its exigencies. Thus, unlike say farmers, they cannot themselves reproduce their means of subsistence.[13] Hunter–gatherers have a mobile and flexible form of social life based on a communal appropriation of the products of labour and the absence of accumulated surpluses or, indeed, of private ownership of productive resources, which lends weight to Engels's claim that before the breakup of society into classes and the rise of the state there existed a stage of primitive communism.[14]

In this egalitarian form of society there are sharp sanctions which are applied against any displays of self-aggrandisement and acquisitiveness which might profoundly threaten the delicate fabric of reciprocal ties and obligations that bind together the collectivity. Lee reports an example of the kind of cultural practices that maintain equality in the group. On the part of a successful hunter the required demeanour of modesty and understatement is described by the !Kung in these words:

> Say that a man has been hunting. He must not come home and announce like a braggart, 'I have killed a big one in the bush!' He must first sit down in silence until I or someone else comes up to

> his fire and asks, 'What did you see today?' He replies quietly, 'Ah, I'm no good for hunting. I saw nothing at all – maybe just a tiny one.' Then I smile to myself because I now know he has killed something big.[15]

If a hunter has had a particularly good series of kills he may even voluntarily cease hunting for a matter of weeks so as not to outshine too far his fellow-hunters. On the other hand, if a hunter's luck seems to have failed him he will be provided with food by his kin until he resumes his activities.[16] The egalitarianism and non-competitiveness of the !Kung is further exposed in the practices of sharing arrows and therefore kills and in the use of hunting magic in which one man will open the tattoos of another in order to reinforce his powers.[17] In the absence of real control over nature, magic and divination offer themselves as substitutes. It is significant in this respect that Lee includes oracle discs among the hunting tools although strictly speaking these are purely illusory means of acting on the real world.[18]

The key to the success of the !Kung way of life does not lie in the complexity of their tools but in their socially interconnected behaviour. So far as hunting is concerned the basic technology is relatively simple. What is genuinely complex is the shared knowledge of the environment and the behaviour of the species it contains. There are, for example, 262 species of mammals, birds, reptiles and insects known to the !Kung of which 80 are eaten.[19] The hunting of larger game may involve the active co-operation of several individuals in tracking, stalking, wounding and killing an animal. Similarly driving animals into a trap is simple in technology but complex in co-operation. Interestingly, Mary Leakey has suggested from the discovery of skeletons of an elephant and other animals embedded in clay and associated with stone tools, that these animals may have been deliberately driven into swamps by early hominids.[20] Among the !Kung, discussions of what and where to hunt 'may occupy several hundred hours a year of a man's time'.[21] Previous experiences are gone over and even men who are not hunting contribute to the discussions. In this way the !Kung attempt to establish control over nature through pooling their accumulated knowledge and skills which become a kind of collective inheritance of the entire group.

In fact, Lee makes the point, that such discussions include women also. Although women do not participate in the actual process of hunting their knowledge of the environment gained on gathering trips is valued by the men.[22] It is generally argued that the hunting of large game with its attendant rigours would have led to the separation of tasks between men and women, particularly as the prolonged care of infants became an important feature in human evolution. It is undoubtedly true that such writers as Washburn and Lancaster, Laughlin and others have tended to view the specifically male involvement in hunting as providing the main stimulus in human development.[23] However, although among the overwhelming majority of known hunter–gatherer groups there is a sexual division of labour between men and women, with the hunting of large game the exclusive province of the former, this may well have been a fairly late evolutionary development.

Several writers have recently pointed to the undervaluation of women's contribution to the evolution of humanity.[24] In particular Slocum argues that the shift from primate individual gathering to human food-sharing cannot be explained simply by hunting.[25] Slocum reinstates the centrality of the labour performed by women in the overall productive process.

> Among modern hunter–gatherers, even in the marginal environments where most live, the females can usually gather enough to support themselves and their families. In these groups gathering provides the major portion of the diet, and there is no reason to assume that this was not also the case in the Pliocene or early Pleistocene. In the modern groups women and children both gather and hunt small animals, though they usually do not go on the longer hunts. So, we can assume a group of evolving protohominids, gathering and perhaps beginning to hunt small animals, with the mothers gathering quite efficiently for themselves and their offspring.[26]

Notions of the male as the socially responsible provider for specific dependants are somewhat suspect in this view. Slocum suggests that the primary locus of food-sharing and the family developed from the mother–child relationship. The techniques of hunting large animals were probably much later developments, after the mother–-children family pattern had been established, and, indeed, it is quite possible that the hunter would share food in the first instance not with his sexual partner but with his mother and siblings.[27] Nor was it necessarily simply co-operative hunting among males which required more skill in social organisation and communication and therefore provided the selection pressures for increased brain size.

> The need to organise for feeding after weaning, learning to handle the more complex social–emotional bonds that were developing, the new skills and cultural incentives surrounding more extensive gathering – all would demand larger brains. Too much attention has been given to the skills required by hunting, and too little to the skills required for gathering and the raising of dependent young. The techniques required for efficient gathering include location and identification of plant varieties, seasonal and geographical knowledge, containers for carrying the food, and tools for its preparation. Among modern hunting–gathering groups this knowledge is an extremely complex, well-developed, and important part of their cultural equipment. Caring for a curious, energetic, but still dependent human infant is difficult and demanding. Not only must the infant be watched, it must be taught the customs, dangers, and knowledge of its group. For the early hominids, as their cultural equipment and symbolic communication increased, the job of training the young would demand more skill. Selection pressure for better brains came from many directions.[28]

Slocum's thesis that there would have been intervening changes before the development of big-game hunting is entirely plausible. Data from the !Kung show that gathered foods provide 71 per cent of the calories consumed, and, indeed, gathering was 67 per cent more productive per person-hour than hunting.[29] As with animal foods, the !Kung had detailed knowledge of about 200 species of plant, in addition to the main staple of the mongongo nut, all of which were named by the !Kung.[30] It has already been suggested that wooden digging sticks may have been the earliest human tools. Certainly these are employed by the !Kung in the collecting of food, as are various forms of containers – useful not only for the transportation of food but also for carrying offspring on foraging expeditions. Like digging sticks, these containers – made perhaps of animal hide, bark or leaves – are perishable in nature and would not have turned up in the archaeological record of early humans. However, as Lee points out, such carrying devices may be seen as an 'essential prerequisite' of the human economy which 'vastly improves the efficiency of labour'.[31] The fact that, in the fossil record, stone tools most often associated with the consumption of meat have been preserved, while the equally important and perhaps historically prior tools associated with gathering and preparation of plant foods have not been preserved, has probably led to an overemphasis on the role of hunting in the evolutionary process.

While the gathering of plant foods may well have preceded the development of organised hunting as a necessary prior stage it is difficult to regard scavenging for animal flesh in the same light.[32] Isaac, for example, seems to suggest a kind of generalised hominid 'opportunism' in which meat scavenging was an important part of their adaptive strategy.[33] However, as both Washburn and Lee point out, as well as being an extremely uncertain method of ensuring a supply of meat with any regularity at all, it would probably involve direct and possibly extremely hazardous confrontation with larger predators and, as such, a very inefficient employment of labour time.[34] Clearly, as Engels hinted, a proportion of meat in the diet was extremely important in enabling adaptation to more varied climatic conditions. As Bicchieri has pointed out, the chances of productive failure, everything else being equal, are halved by a mixed hunting–gathering economy. 'Practising such mixed exploitation a group can, more often than not, fall back on the alternative technique when necessary.'[35] Whether the consumption of meat produced chemical substances in the human brain that were 'premises' for the transition to humanity, as Engels thought, is more doubtful. It was probably the high quality nutrients relative to vegetable foods which were valued.[36] What seems likely is that early hominids increasingly became more carnivorous than their nearest anthropoid relatives.

This fact has been used by Raymond Dart and his followers to extrapolate in the reverse direction, from early man to contemporary *Homo sapiens.* In an essay, the title of which probably unconsciously parodies that of Engels's, Dart examines 'The Predatory Transition from Ape to Man'. Dart claims, 'The loathsome cruelty of mankind to man forms one of his inescapable, characteristic and differentiative features; and it is explicable only in terms of his carnivorous and cannibalistic origin.'[37] The views of early man as a killer-ape, and, by implication, of modern man also as an innately aggressive species, have of course been given a weighty ideological if not scientific significance by Dart's admirer and populariser, Robert Ardrey.[38]

For Dart also it is clearly hunting by the male of the species which has provided the main impetus for the development of human beings. Dart argues that the South African australopithecines whose remains he has uncovered in a number of cave locations had developed what he termed an '*osteodontokeratic* culture', that is a culture based on tools and weapons made of bone, teeth and horn.[39] These australopithecines purportedly used the long limb bones of antelopes as clubs, teeth extracted from carnivores as daggers and horns as primitive skewers or gouges.[40] Not only did they hunt other animals but also, on occasion, members of their own species. Commenting on the various skeletal remains found along with those of early man in South African caves, Dart observes,

> these large bones are not gnawed but split and crushed to extract the marrow; and the double-ridged extremities of the arm bones (humeri) of antelope are cracked and fractured through having been used as bludgeons. The broken ridges of these humeri correspond with double-furrowed fractures found in baboon (and even in australopithecine) skulls found at all three man-ape sites: Taungs, Sterkfontein and Makapansgat.[41]

Dart had found serious disproportions in the parts of skeletons preserved as fossils and argued that the missing bones were simply not brought back to the cave at all by the resident australopithecines. This view has been contested by Brain in an important study which examined the preservation of skeletal remains of goats consumed by the Kuiseb River Hottentots alongside the Makapansgat assemblages. Brain concludes after careful comparison that, 'the overall similarity in composition of the bone collections is remarkable. It is a reflection of the predictable pattern of survival which manifests itself when whole skeletons are subject to destructive treatment.'[42]

Brain argues on the basis of the evidence of the Kuiseb River goat bones that it is unnecessary to postulate artificial selection of skeletal parts to explain the Makapansgat remains. The more resistant portions of limb bones survive with greater frequency due to a physio-chemical fusion process, with the result that the distal ends of humeri survived to the virtual exclusion of proximal ends.[43] The status of Dart's armoury of primitive 'clubs' is clearly called into question.

A similar line of attack is pursued by Feustel in a critical comment upon Wolberg's attempt to vindicate Dart. Feustel takes exception to Dart's contention that these remains are tools or represented the products of a definite cultural tradition.

> Many of these objects show forms that are implied by their specific bone structures, forms that could be the same whether the force that produced them was an animal's teeth, the pressure of the earth, a blow from a stone, or the intentional act of man. Furthermore, no matter what their origin, fragments of bone and antler are all transformed in the same way during their resting in the ground, where physical and chemical influences caused smoothing, polishing and edge-rounding.[44]

Whether or not australopithecines actually *lived* in these caves is also a matter of dispute since they would be in competition with other predators who may well have had their homes there and, lacking fire as a means of protection, hominids would have been relatively vulnerable to attack. Some interesting related evidence on this question has been gathered by Brain.

In an ingenious analysis of the fossil remains in the Swartkrans cave deposit, Brain shows that such remains may well have come from sources which do not imply human dwelling at all.[45] The original cave entrance led by a descending shaft to a subterranean cavern. It is in this latter area that the fossilised bone accumulations are found, the former entrance and connecting shaft having been long since eroded. Brain suggests that the large trees likely to have grown around the original cave entrance, where they would be sheltered from damage by frost or fire, could have provided storage larders for leopard kills, out of reach of scavenging hyenas. Over possibly 20,000 years or so it is not unlikely that substantial amounts of bone might fall off the trees as leopards consumed their victims. 'Food remains of leopards are consequently introduced into the catchment areas of the cave entrances and some will ultimately find their way down to the fossilisation sites.'[46]

The large number of bone fragments are identifiable as belonging to *Australopithecus* together with remains of antelopes, baboons and hydraxes. Interestingly, it is overwhelmingly cranial parts of *Australopithecus* rather than other more easily broken parts of the skeleton which remain – likewise suggesting possible consumption by leopards. In this connection Ardrey's claim that one of the hominid specimens was killed by being struck twice on the back of the head by a pointed object is thrown into doubt by Brain's study.[47] Brain shows that it is, 'more likely that the two holes were made simultaneously by the canine teeth of a carnivore spaced about 33mm apart.'[48] Similar holes in the skulls of baboons eaten by leopards have been reported. It seems more likely that the characteristic mode of dragging the body with the head of the victim clamped between the leopard's teeth is responsible for the holes in the *Australopithecus* skull rather than any prehistoric act of murder by a fellow-australopithecine. As Brain points out, leopard predation still occurs today and may have been much more prevalent at an earlier stage of human evolution, when neither stature nor defensive weapons were as well developed.[49] Although the main contention of Dart's hypothesis – that there was an actual cultural tradition of bone, teeth and horn implements – lacks plausibility, in this writer's opinion, it may be that along with wood and stone, animal bones and remains were fashioned into tools. This, however, is a long way from suggesting that predation and aggression were the characteristic modes which determined the human line of advance. In his attempt to portray an essentially Hobbesian view of mankind, Dart substantially underplays the socially co-operative dimension of labour activities.

Here again, although contemporary hunter–gatherers are in fact much more advanced than australopithecines, a wide range of studies suggest a life-style based upon peaceful co-existence and harmony rather than aggression. Lee, Woodburn and Turnbull among others suggest that aggression, if not entirely absent in hunter–gatherer society, has a very secondary role in their affairs.[50] Indeed, it could be argued that early humans, living within the 'primitive herd', an essentially primordial form of social association, would have been unable to build more enduring human collectives unless unbridled 'zoological egoism' began to be subordinated to the type of sustained social co-operation required for successful labour.[51] The hunting-and-gathering way of life provided the social framework within which was secured the first embryonic cultural attainments in the form of tool-making, food-sharing, the prolonged socialisation of the young and perhaps the rudimentary division of labour. These manifest advantages of social co-operation pointed in the broad general direction of humanity's further social advance.

4 Primate tool-use and tool-making

From primatology a variety of works have been forthcoming which are claimed to have a bearing on the evolution of early man.[1] Studies of mankind's nearest primate relatives have been used to highlight certain aspects of human behaviour. As with ethnographic material, however, these comparisons must be carefully drawn as the subsequent discussion is intended to show. First, some insight into the question of hunting can be gained by examining the behaviour of anthropoids who occupy an environment not all that dissimilar to that which early humans may have faced. Baboons, for example, are ground-living primates subsisting mainly on vegetable-matter with only 1 per cent of their diet being provided by animal protein.[2] Most baboon kills described by Harding were of small antelopes and hares, animals which defend themselves by 'freezing', that is by remaining immobile. There was no evidence of co-operation between baboons during hunting, and females were never allowed to keep their catch.[3] Moreover, among the males there was no voluntary sharing – indeed, the killing and eating of prey generally caused a marked rise in the level of aggression in the troop.[4]

De Vore and Washburn observed that baboons made no attempt to eat fresh carrion when it was found, and that scavenging plays no part in the consumption of meat.[5] Further, nothing resembling the stalking of prey was ever observed and, as such, baboons may be described as 'very inefficient predators'.[6] Even so far as the grasses which comprise the great bulk of baboons' food supply were concerned, during the dry season when it is necessary to dig underneath the hard soil to extract moisture-bearing roots, no baboon was observed trying to use a tool in this or any other way. As De Vore and Washburn conclude:

> The use of a stick or a stone for digging would increase

> the baboon's food supply more than any other simple invention. Perhaps in *Australopithecus* we see a form which had such a tool to exploit vegetable foods and which also used this tool as a weapon.[7]

Chimpanzees also consume a limited amount of meat. However, the evidence suggests that such meat-eating occurs in cycles. Chimpanzees are not always 'on the hunt' but rather as Suzuki says, 'the hunting and meat-eating of chimpanzees seems to be related to some socially excited situation in their nomadic life that is dependent on changes in their food seasons.'[8] Much of the time chimpanzees will ignore potential prey within easy reach.[9] Interestingly, Suzuki notes that the chimpanzees also do not feed on carrion.[10] Unlike baboons, however, there exists certain evidence of what appears to be co-operative hunting and food-sharing among chimpanzees.[11] Usually chimpanzees do not search for or pursue their prey. The most common prey of chimpanzees, young baboons, are the victims of 'simple seizure' as Teleki describes it, that is, a sudden lunge followed by 'instaneous capture' by a single adult.[12] Stalking behaviour was notably unsuccessful, while chasing the prey was only slightly more successful – but in neither baboon nor chimpanzee does persistence in the hunt appear to be well developed.[13]

Goodall has described a hunt which involved chimpanzees in purportedly co-operative activity, the prey being a juvenile baboon. While one chimpanzee climbed a tree to approach it, other chimpanzees stood below both that tree and other trees which might act as escape routes for the quarry.[14] Impressive though this appears, the fact remains that chimpanzees are essentially opportunistic meat-eaters.[15] Moreover, other evidence of co-operative hunting among chimpanzees is open to alternative interpretations. The frequent shift in the leading position from one chimpanzee to another, regardless of the social rank of the individual, during a chase described by Teleki, could be held to suggest not so much flexible co-operation as a more random mêlée-type of behaviour, well below the threshold of what could be regarded as socially co-operative.[16] Indeed Kortlandt suggests that the so-called predatory behaviour of great apes is, 'actually to a large extent intimidation behaviour, that is showing off toward insiders, redirected toward outsiders, and at the same time, extermination of food competitors'.[17] While chimpanzees will eat young baboons, monkeys and bushbucks – which are all food competitors – they do not consume other small to medium sized vertebrates which are easily accessible. Co-operation in active search for prey seems confined to humans.[18]

With regard to food-sharing among chimpanzees, so far as the division of prey is concerned, this appears only slightly less limited in scope than among baboons. Teleki has described in detail how, initially after capture, the chimpanzee who got there in time could try to take meat without interference, irrespective of dominance ranking.[19] They then became 'proprietors', as it were, of whatever part of the kill they had managed to 'hoard' and subsequent newcomers did not attempt to grab at any part held.[20] There was no marked increase in the level of intra-group aggression. Rather they began to form 'sharing clusters' and would beg meat from the possessor, a behaviour which Isaac has described as 'tolerated scrounging'.[21] Interestingly, individuals could wait in vain for a lengthy period, and, moreover, high social rank was no guarantee of success in requesting meat.[22] Further, Isaac comments on the total absence of co-operative behaviour as far as the chimpanzee's main food is concerned, 'Vegetable foods, which are the great apes' principal diet, are not shared and are almost invariably consumed by each individual on the spot.'[23] There is, then, no regular postponement of food consumption until they have returned to a home base nor do they actively seek out and procure prey or share food in the manner it has been suggested humans have done since their origins.[24]

While the hunting techniques of early humans can only be guessed at, in one important respect at least, their behaviour was probably marked by a degree of foresight and purposiveness far greater than their anthropoid relatives, if contemporary primatology is anything to go by. No chimpanzee has been observed employing weapons as hunting aids in the killing of live mammal prey, nor have any cutting or butchering tools been seen to be used to divide up the carcass.[25] Chimpanzees in fact kill either by biting the back of the prey's neck, by bashing out its brains or by twisting the prey's neck with both hands.[26] Baboons make no attempt to skin their prey, the meat in general being eaten from the inside.[27]

Nevertheless, the remarkable studies of free-living chimpanzees in the Gombe National Park of Jane Goodall and her colleagues have shown a level of tool-using and even tool-making activity which had formerly been regarded as only capable of being performed by humans. Goodall identifies two types of activity in which tool-use occurs: food-gathering and emotional or 'agonistic' displays.

> At Gombe many instances of tool use occur in a feeding context: grass stems or thin twigs are used to 'fish' termites from their nests; sticks are pushed into underground or arboreal nests of various species of ants; sticks may also be used as levers to enlarge the openings of underground bees' nests. A leaf 'sponge' ... is used for scooping up water in the hollow of a tree which the chimpanzee cannot reach with his lips, and a sponge of this sort was once used for cleaning out the inside of a baboon skull after a kill. Twigs and sticks are also used as investigation probes, again sometimes in a feeding context. Thus a twig may be inserted into a hole in dead wood: after sniffing the end the chimpanzee either discards the wood, or presumably in response to some olfactory cue, tears it apart and (usually) finds and consumes a beetle or wasp larva. Frequently a number of openings in a termite nest are investigated thusly before the chimpanzee begins work in earnest.[28]

Goodall has also related that leaves may be used to wipe dirt or blood from the body, as 'toilet' aids, and that branches may be waved during agonistic displays and occasionally sticks or stones were aimed and hurled both underarm and overarm, sometimes playfully, but on a few occasions as part of threatening displays.[29] Interestingly, when a real fight begins, chimpanzees drop the sticks they may be holding and rely on teeth and hands much as they do when hunting. As Lancaster has suggested, however, this single population of chimpanzees at Gombe has performed more complex kinds of tool-use, in a wider variety of situations, than has been observed for any other animal and this despite the fact that tool-use is a very small part of their behavioural repertoire and is a comparatively rare event.[30]

What comprises the difference, if any, between the employment of tools by chimpanzees and *human labour* is the key issue which must be resolved. Here a real measure of the character of the activity implied by chimpanzee tool-behaviour can only be gained by examining the more complex question of tool-*making*. On this crucial question Goodall provides some surprising information on chimpanzee life.

> In addition to using the objects around them in the above ways, the chimpanzees sometimes modify the material to make it more suitable for the purpose in hand. Thus, during termite fishing, leaves may be stripped from a twig, a strip of bark may be shredded, or the blades may be stripped from a wide length of grass. When drinking from a hollow, the chimpanzee almost always crumples the leaves by chewing them briefly before using them as a sponge, thus increasing their water-carrying capacity considerably. These simple modifications may be considered the primitive beginnings of tool-making.[31]

The continuities between chimpanzee tool-behaviour and that of humans are clearly traced by Goodall. In this respect, the discontinuities – what chimpanzees cannot do – assume particular importance in accurately identifying the unique character of human labour activity.

In a useful review of Goodall's evidence, Guilmet notes that all of the reported chimpanzee tool-making behaviour is directed toward the modification of easily shaped materials such as stems, stalks, sticks and leaves.[32] Moreover, such tool-making consists of the simple modification of soft materials using only the natural organs of the body such as teeth, lips and fingers.[33] In this respect, a series of experiments conducted with chimpanzees by the Soviet investigator, Khrustov is of central relevance.[34] Khrustov conducted two sets of experiments with the chimpanzee Sultan and two control chimpanzees. In the first of these, the experimental animal succeeded in fashioning a tool with definite parameters from a material of indeterminate shape, in this case by pulling off a sufficiently narrow strip of wood from a wooden disc which could then be inserted in a metal pipe in order to retrieve a lure. This showed chimpanzee tool-making behaviour on a par with that observed by Goodall among free-living chimpanzees.

In the second series the material to be fashioned was superficially similar in form except that the disc was made of harder oak wood and as such, a strip could only be cut off with the aid of an auxiliary tool. In other words, the chimpanzees had not simply to make the tool, but to use another tool in order to make the tool. In these experiments the chimpanzees were given a Chellean hand axe, historically designed specifically for similar operations. Despite the chimpanzees' manifesting considerable ingenuity in their actions, trying stubbornly in different ways to fashion the required tool, '*none of them* attempted to fashion the unyielding material with the stone tool offered along with it, or with any object other than *their* own natural body organs.'[35] Even when a repeated and complete demonstration of the entire operation and ensuing finished tool, made with the aid of the stone axe was presented, in no case did they attempt to imitate the action of the experimenter and use the tool provided.[36] Khrustov's work, posed with the question in mind of the possible implications for human origins, would seem to suggest that the use of a tool to make another tool involves a kind of foresight and purposiveness which takes mankind beyond the 'highest implemental frontier' of the apes.[37]

It is perhaps significant in this respect that Khrustov conducted his experiments using a Chellean hand axe, a stage of tool-making development which archaeologists have identified as coming later than that of the australopithecines. Guilmet points out, however, commenting on the dangers of undervaluing the Olduvai evidence regarding tool-making:

> By the appearance of the first stone tools in the archaeological record the hominid capacity for object modification and use has surpassed the tool-using and tool-making capacities of living nonhuman primates in the wild. The tool kit of the australopithecines appears much more diverse than that of any living nonhuman primate indeed, of all living nonhuman primate species in the wild.[38]

As Guilmet suggests, if each stone tool had at least one unique function, then australopithecines had nineteen behaviour patterns related to tool-use and nineteen behaviour patterns related to tool-making and this is a conservative estimate, since it excludes non-stone tools and the fact that early tools were relatively unspecialised and may have been used for multiple tasks.[39]

Oakley has pointed out that even the oldest industries of the Oldowan type already contain *tools for the making of other tools.*[40] Such tools as hammerstones, for example, have no immediate object in themselves but are employed as intermediate tools in a preparatory manufacturing operation. The use of a tool to make a tool suggests a reallocation of social energies from immediate episodic subsistence to general preparation for future subsistence activity. In other words, it indicates the beginning of *planned social production.* It is this that distinguishes the emergence of human tool-activity from that of non-human primates. Gruber comments on this point,

> Oakley's definition of 'man the tool-maker' holds, if we add the criterion of the modification of a natural object, *using another object or some implement as an aid in manufacture.* This adds the notion of the utilisation of an intermediary device as an aid to create a true (human) tool, i.e. another object or some implement in the manufacturing process, which can range all the way from a completely unworked stone (e.g. a pebble with a sharp edge or a hammerstone) to a sophisticated complex modern implement. Nonhuman primates in the wild have never been shown to have this ability.[41]

It could be argued that Khrustov ought to have conducted his experiments with the much earlier technology of pebble tools rather than the hand axe. This would more nearly correspond to what is now known of the earliest forms of human labour. In this respect the less methodologically sophisticated experiments with a young orangutang at Bristol Zoo conducted by Wright are of some interest.[42] The aim of these experiments was to get an ape, after repeated demonstrations to use a hammerstone to strike a flake from a flint core, and to use this flake as a tool with which to cut a cord, thereby opening a box and retrieving food.[43] Significantly, Wright makes no claim for the degree of insight implied by the successful carrying out of this task by the orangutang.[44] The results do suggest though that the employment of auxiliary means in order to manufacture a tool may be potentially *within the capabilities* of an anthropoid after all, especially if the necessary conditions for this behaviour are experimentally created.[45] It must be stressed that as yet no observations have been recorded of such highly 'mediated' forms of tool-making activity among free-living primates. As such, the evolutionary implications of Khrustov's main conclusions would still seem to apply.

Moreover, it is open to question as to whether the type of learning process Wright described goes much beyond the kind of trial-and-error learning commonly found in practically all living vertebrates.[46] In a review of a wide range of animal tool-behaviour – including such diverse creatures as wasps, crabs, birds, otters, bears, elephants as well as anthropoids – Hall argues that none of these is necessarily of significance in the evolution of 'intelligent' tool-using. The fact that a behavioural rather than a physical adaptation has taken place to deal with some ecological condition does not necessarily indicate a more complex type of learning.[47]

> These adaptations do not appear to give their possessors any selective advantage over other species which have evolved alternative forms of adaptation. Rather, they simply enable their possessors to survive at a certain population level in their ecological niches. In other words, such performances are only worthy of special note because of their entirely superficial, indeed one might almost say fortuitous resemblance to human tool-using.[48]

More recently Beck has argued also that there is no orderly relationship between tool-behaviour and intelligence which would necessitate the development of special cognitive sophistication in the utilisation of tools by animals.[49] Beck concedes only a very limited role for observational learning in the acquisition of animal tool-behaviour and is uncertain as to whether any instances of insightful learning can be substantiated.[50] It is argued, in an attempt to downgrade the specificity of human labour, that tool-behaviour in animals is mainly an adaptation to 'extractive foraging on embedded foods' and in this sense can be said to provide an evolutionary parallel to the rudimentary tool-activity of early humans in extracting roots from the soil, insects from nests or the edible interiors of plant and animal foods.[51]

In a similar vein, the fairly limited level of tool-using and tool-making observed among chimpanzees in the wild is regarded as sufficiently close to that of humans for it to be suggested that chimpanzees are capable of intra-species *cultural transmission*.[52] Goodall, for example, says,

> It therefore seems probable that the use of sticks, stems and leaves for the specific purposes described here represents a series of primitive cultural traditions passed on from one generation to the next in the Gombe Stream area.[53]

Further studies by McGrew and Tutin at two locations in Africa, on variations in social-grooming behaviour and on differing forms of tools employed in termiting at three locations, have attempted to specify more closely the contours of chimpanzee 'culture'.[54] In their study of termiting McGrew *et al.* identify three types of behaviour variations which they argue cannot be put down to results of environmental constraints. There are differences in selection of tool-materials, for example, use of twigs and leaf-stalk as against grass and bark, differences in how much bark is peeled from the probes, and differences between chimpanzee populations as to whether one or both ends of the probes are used.[55]

The absence of a human level of foresight and purposiveness in regard to chimpanzee tool-making has already been mentioned. Even with respect to tool-using by chimpanzees the predominant pattern in termiting expeditions, for example, is for the raw material of the tools to be collected if possible near the site of food-gathering. Pre-selection of tool material is limited. While the tool may be retained for some considerable time and carried some distance, once the immediate search is ended, as the large piles of twigs and stems left at termite mounds attest, the tool is discarded after its use.[56] This can only in part be explained because of the generally destructible nature of the materials. As Guilmet points out, by contrast, 'Australopithecines most likely saved carefully shaped stone tools for later use indicating the recognition that it would be more advantageous to save them rather than to make new ones when the immediate need arose.'[57] In saving the tools they were, in effect, also storing up the objectified experience of previous labour activity. For developing humans the possibility existed of eventually modifying and improving their tools in the future, thus increasing the efficiency of their labour activity. To the degree to which this was done, as labour became more complex each succeeding generation would become more genuinely 'tool-dependent' – provided there existed some means for the handing on of such experience.[58]

White, in a pioneering article, sums up why apes cannot be said to possess even a primitive cultural tradition.[59]

> The fundamental difference lies in the fact that the use of tools among man is a cumulative and progressive process whereas among apes it is neither. This is not to say that an individual ape does not make progress in his use of tools nor that he cannot increase his repertory of tool behaviour. What we are saying is

> that apes as species make no progress in tool-using, one generation is no further advanced than its predecessor.[60]

Even McGrew and Tutin are forced to admit to the fact that chimpanzees are 'culturally conservative'.[61] With humans, on the other hand, the reverse is the case. Each generation can add to and build upon the tools and techniques of preceding generations although as has been pointed out, this at first was probably an immensely long drawn-out process. For White, it is the use of the symbol, particularly in *word* form which makes the tool-experience of mankind both subjectively and objectively a continuous and enduring cultural phenomenon.[62] It is suggested by White that the development of some form of language as the repository of such tool-experience and the storehouse of socially useful information to be passed on to succeeding generations, guarantees the cumulative and progressive character of human culture.[63] It is this which takes human tool-activity beyond the 'highest implemental frontier' described by Khrustov. The question of language and culture is examined further in the following chapters since it is crucial to an understanding of the key discontinuities between humans and other animals.

5 Primate communication and culture

If anthropoids could be taught to master human language then it would appear that, as Mounin has put it, 'the ultimate defensive gap' in the debate over the radical separation between man and ape has been bridged.[1] The significance of such attempts to communicate with anthropoids through the use of human language is crucial to an understanding of wherein, if anywhere, lies the uniqueness of mankind. From the pioneering work of the Hayes with the chimpanzee Vicki at the Yerkes Laboratories of Primate Biology during the period 1947–54 it was established that *spoken* language was not a suitable medium in which to establish 'two-way communication' with a chimpanzee. As a result of that failure to teach Vicki to speak more than seven words up to the time of her death, the Hayes suggested that future researchers might experiment with the use of physical gestures as an alternative medium.[2]

This advice has been taken up by the Gardners who raised the chimpanzee Washoe in a 'stimulus-rich environment' with a team of assistants using a gestural system of communication called American Sign Language (ASL).[3] After thirty-six months Washoe was reported as reliably using eighty-five signs.[4] Moreover, Washoe employed numerous multi-sign combinations even producing new combinations of her own.[5] While the Gardners were reticent about whether or not Washoe could actually be said to possess a degree of mastery of human language there was a clear implication that such a possibility could no longer be ruled out.

In a subsequent study by Premack with the chimpanzee Sarah, this possibility was further explored.[6] The medium of communication employed this time consisted of plastic counters varying in shape, size, texture and colour which could be adhered to a magnetised board. Premack claimed that using this apparatus over eighteen months he was able to train Sarah to use and understand the negative particle, *wh*-questions, the concept of 'name of', dimensional classes, prepositions, hierarchically organised sentences, and the conditional. It appeared, in Premack's view, therefore, that mastery of the rudiments of grammatical forms equivalent at the very least to that of a human child aged 2½–3 years was possible.[7] These findings would seem to be supported by the work of Rumbaugh and his colleagues using a computerised keyboard system of 'lexigrams' with the chimpanzee Lana. The use of the computer as an interactive device in this project was in order to rule out the 'Clever Hans' phenomenon, that is, the possibility that non-linguistic cues might be given by the chimpanzees' testers which could covertly shape their responses as a result of these experiments. Rumbaugh has suggested most forcefully that language can no longer be viewed as a uniquely human characteristic.[8] This would also seem to be the view of Washoe's subsequent guardian, Roger Fouts who has claimed that chimpanzees may even use ASL to communicate with one another in such situations as mutual comforting, eating and general play activities although, significantly, no claim has yet been advanced that chimpanzees can transmit ASL to their offspring.[9]

Of great interest, therefore, are the careful observations of the chimpanzee Nim reported by Terrace and his associates.[10] Over 19,000 multi-sign utterances in ASL made by Nim were analysed for syntactic and semantic regularities. Terrace *et al.*, pointed out that the Gardners did not report the order of signs of Washoe's multi-sign combinations and that therefore there was no basis for deciding whether Washoe was obeying the rules of sign-order.[11] Nim's linear combinations of signs were subjected to three forms of analysis. First, distributional regularities in two-sign utterances were examined to establish whether or not lexical regularities existed. Second, having established that such lexical regularities did exist, semantic relationships were investigated. Third, a 'discourse' analysis of videotape transcripts between Nim and his teachers was undertaken.

Terrace *et al.* argued that a large variety of combinations, greater than could be retained by memory alone, is still not sufficient to demonstrate that such combinations are sentences expressing a semantic proposition in a rule-governed sequence of signs. The regularity of Nim's sign-combinations did suggest they were generated by linguistic rules. For example, *more* + *X* is more frequent than *X* + *more.* Such frequency patterns, however, need not have been structurally constrained but could have been the result of independent first- and second-position habits rather than syntactic rules.[12] On the basis of observed distributional regularities an attempt was made to predict the frequencies of different two-sign constructions by multiplying the probabilities of the relevant lexical types appearing in first and second positions respectively. The correlation between predicted and observed probabilities was not statistically significant suggesting Nim's two-sign sequences were *not* formed by independent position habits – that is, that some form of linguistic construction was taking place.[13]

Next, comparisons were made between Nim's two-, three- and four-sign combinations. Nim's longer utterences, unlike those of a child, were not semantic or syntactical elaborations of his shorter utterances. As children increase the length of their utterances they elaborate their initially short utterances to provide additional information about some topic. Nim, on the other hand, merely tended to repeat the same lexical items in his three- and four-sign combinations not so much to add new information but instead to add emphasis.[14] Moreover, as the mean length of a child's utterances (MLU) increases, its complexity also increases. By contrast Nim's MLUs during a 19-month period showed a marked failure to grow.

With regard to the aforementioned studies of an ape's purported ability to express semantic relationships in combinations of signs, Terrace and his colleagues argue that these have yet to advance beyond the stage of 'unvalidated interpretation'.[15] They instance Fouts's widely cited report that Washoe was able to create new meanings through novel combinations of signs as illustrated by Washoe's signing 'water bird' when confronted by a duck in a lake. This was a kind of bird Washoe had never before encountered.[16] Terrace suggests rather that no matter how compelling anecdotal evidence may seem,

> Washoe may have simply been answering the question, *what that?*, by identifying correctly a body of water and a bird, in that order. Before concluding that Washoe was relating the sign *water* to the sign *bird,* one must know whether she regularly placed an adjective *(water)* before, or after, a noun *(bird)*.[17]

The semantic aspect of Nim's combinations were further examined through discourse analysis of video transcripts. These revealed that Nim's utterances were often initiated by his teacher's signing and were often full or partial imitations of the teacher's preceding utterance. Adjacent utterances are those that follow an adult utterance without a definitive pause. A high percentage of Nim's utterances were classified as adjacent, falling into one of four categories: imitations, containing all the adult's lexical items; reductions, containing some of the adult's lexical items and nothing else; expansions, containing some of the adult's lexical items and some new ones; and novel utterances, containing none of the adult's lexical items.[18] Imitations and reductions decrease with increasing mean length of utterances for children. By contrast a far greater proportion of Nim's adjacent utterances were imitations or reductions. On the other hand, as compared to young children a far smaller proportion were expansions. Whereas expansions increase as a child gets older, particularly systematic expansions of verb relations contained in the adult's prior utterance, no such pattern was discernible in Nim's utterances, which contained only a small number of additional signs and did not add new information to the teacher's utterance.[19]

By definition, adjacent utterances may include interruptions of a teacher's or adult's utterance. This results in simultaneous rather than successive discourse. With Nim, the turn-taking and dialogic character of human speech was mainly absent. Video analysis revealed that in 71 per cent of the utterances examined, Nim signed simultaneously with his teacher, and of these over-lapping utterances 70 per cent occurred when Nim began an utterance while the teacher was signing.[20] None of Nim's teachers, nor the expert observers who were fluent in ASL, was aware of the extent to which this had been the case, prior to seeing the videotape analysis. Even more telling was the fact that a similar analysis of films of Washoe's signing revealed a comparable phenomenon.[21] The writers suggest that far from chimpanzees producing and understanding sentences, previous investigators have failed to rule out alternative explanations of apes' 'signing' behaviour. For example, they suggest the possibility of rote learning of sequences of signs of what, for the ape, are really nothing but nonsense symbols, except for the names of the reward objects requested. Even demonstrations that chimpanzees can answer *wh*-questions correctly which would seem to require an understanding of syntactic structure, could as easily be explained as the result of the establishment of non-syntactic learning sets in constant settings.[22] Terrace and his co-workers conclude:

> In sum, evidence that apes create sentences can, in each case, be explained by reference to simpler nonlinguistic processes. Sequences of signs, produced by Nim and by other apes, may resemble superficially the first multiword sequences produced by children. But unless alternative explanations of an ape's combinations of signs are eliminated, in particular the habit of partially imitating teachers' recent utterances, there is no reason to regard an ape's multisign utterance as a sentence.[23]

It would appear that laboratory-based attempts to teach chimpanzees to communicate via an accessible form of human language are not well founded,[24] a conclusion which has recently received further support from other investigators.[25] Thompson and Church simulated the language behaviour of the chimpanzee Lana by a computer model in which the animal selects, depending upon context, one of six stock sentences with fixed and variable elements. On the basis of their examination of conversational sequences from the Lana project it was suggested that Lana's use of 'Yerkish' keyboard language could be accounted for by the two basic processes of paired associate learning and conditional discrimination learning. Thompson and Church conclude that given the goal-orientated nature of both Nim and Lana, 'the animals were motivated to produce strings of signs or lexigrams in order to obtain a desired object (or event).'[26]

For primatology perhaps the key questions are those addressed to modes of communication in the wild.[27] The results of this work are of interest in that they point towards the very different nature of primate vocal communication systems from that of humans. Certainly chimpanzees can produce a wide variety of sounds ranging from growling, grunting, barking, panting to screaming and screeching. However, communication among monkeys and apes is not even primarily vocal, relying rather on a multi-modal system of signals of a tactile, visual, olfactory as well as vocal-auditory nature. As Lancaster observes in a review of the field:

> A vocalisation, a gesture, or a facial expression in itself usually does not represent a complete signal, but is only a part of a

> complex constellation of sound, posture, movement, and facial expression. Parts of such a complex pattern may vary independently, and may help to express changes in intensity or level of motivation.[28]

Where group members are within sight of the rest of the group most of the time such proximity allows the complex of multi-modal signals to be easily received and understood. Vocalisation can serve to call visual attention to the signaller or to emphasise or enhance the effect of visual and tactile signals but is not necessarily the primary medium of social communication.[29] For species of primates that live in dense forests, loud vocalisations are commonly used in long-range communication to facilitate group cohesion, signal predators, or achieve intergroup spacing, although such vocalisations are by no means entirely absent in ground-dwelling species.[30]

As Marler has pointed out, primate communication permits subtle changes in mood of the members of the group to be conveyed while providing for co-ordination of its various social activities, particularly those associated with forms of dominance and subordinance relationships.[31] Besides being multi-modal and often graded in form, primate communication is further complicated by the requirement to take into account the social and environmental context of messages.[32] Lancaster has commented that these problems together with the difficulties in analysing visual photos of patterns of movement found in facial expression and gestures make the study of primate communication no easy task.[33] It is clear, though, that in contrast to human language which is capable of conveying a store of more or less objective information between communicants, mainly within a single sensory vocal-auditory modality, multi-modal primate communication is primarily emotional and conveys little in the way of specific information about their physical environment, a contrast which Marler emphasises.

> Environmental information, present or past, figures very little in the communication system of these animals, and a major revolution in information content is still required before the development of a variety of signals specifying certain objects in the environment and a system of grammar to discourse about them can be visualised.[34]

If Marler is correct and human language represents such a 'major revolution' marked by the ability to transmit environmental information, freed from its immediate temporal context, the problem remains of explaining how this radical difference from primate communication came about, an issue which will be returned to in the final chapters. Thus, while under laboratory conditions and in the field certain socially communicative skills have been suggested among anthropoids, it is arguable whether these in fact pre-figure the development of specifically human linguistic abilities. It has also been suggested that the character of anthropoid tool-behaviour can be distinguished from the tool-manufacture and co-operative social labour of humans from an early point in hominid development. Nor is there a basis for claiming that there is any observable interrelated development of tool-making and social communication among anthropoids such as is necessary for the growth of human culture.[35]

It can be argued that only humans possess the essential ingredients of cultural development and that the term 'culture' therefore should be reserved for them alone as a universal distinguishing characteristic. While no attempt is being made artificially to counterpose the cultural and the biological, it is important to retain the specific identity of the various factors which underlie social behaviour if they are to be given the correct explanatory weight. Anthropomorphic descriptions of animal behaviour in terms of human social behaviour which subsequently *derive* those same behaviours from their 'animal origins' necessarily result in the biologisation of culture, the suppression of the unique features of human development. Nowhere is this to be seen more clearly than in the current attempt by sociobiology to refurbish earlier ethological speculations about humanity's 'original nature' derived from the comparative study of animal behaviour with the new and seemingly rigorous science of genetics. Gone are the vulgar assertions of Dart and Ardrey. It is not simply *generic* continuities between human and animal behaviour which are now stressed but irreducible and intractable *genetic* continuities. What were previously suggestive behavioural analogies have been replaced by hard biogenetic *homologies* which direct attention to behaviour patterns purportedly coded in an inherited and shared genetic substratum.[36]

The name of the Harvard biologist E.O. Wilson is most commonly cited as the founder and guiding light of sociobiology.[37] It is not the intention here to conduct a full-scale critique of sociobiology. Thankfully this has already been done elsewhere.[38] Rather it is merely intended to suggest certain problems with this approach in the area which directly infringes on the subject of this discussion and where the theory should have the greatest scientific purchase, namely the question of human origins. Wilson addresses himself to this issue in his popularising 'manifesto' of sociobiology, entitled significantly *On Human Nature.*[39]

In a blend of neo-Darwinian theory and population genetics, sociobiology posits as the determinate process underlying human society the general evolutionary propensity of individual genotypes to maximise their reproductive success, the so-called 'inclusive fitness' of the self-interested gene.[40] Such genetic theories are invoked as the basis of forms of human activity like aggression, sexual behaviour, religious belief, altruism, and so forth, which are characterised as fundamental 'predispositions' – as it were, biologically pre-given. Wilson's work therefore raises in stark form the issue of the interrelation of the biological and the social in human development. In his discussion of the emergence of early humans he proposes a 'dual track' explanation whereby 'genetic evolution by natural selection enlarged the capacity for culture, and culture enhanced the genetic fitness of those who made the maximum use of it.'[41]

If this were all Wilson is saying, as a statement in itself, it would be unexceptionable. As has already been suggested from the work of Washburn and others, it is likely that for early hominids there was an interplay of the biological and the cultural involving a relationship of reciprocal feedback, whereby there was increasing biological selection for a cultural way of life. The biological pole of development, while far from entirely ceasing to operate, became progressively subordinated to the pole of cultural development as early forms of social organisation for labour came into being. In saying that there was increasing selection for the *capacity* for culture, Wilson is doing no more than underlining the obvious fact that the emergence of cultural processes neither obliterates nor should be artificially counterposed to the biological.[42] As Dobzhansky has observed, culture and its genetic basis developed together.[43] In effect, the biological provides the necessary organic preconditions for the development of cultural forms. In his account of the development and characteristics of early hunter–gatherer society as a basically co-operative, peaceful and even egalitarian form of social organisation, Wilson adheres to the orthodoxy, even conceding that 'aggressiveness would have to be restrained and phylogenetically ancient forms of overt primate dominance replaced by complex social skills.'[44] This is not at all the picture which one expects a sociobiologist to present of 'human nature'. All the more surprising, therefore, is the contradictory assertion a few pages later that

> the basic social responses of hunter–gatherers have metamorphosed from relatively modest environmental adaptations into unexpectedly elaborate, even monstrous forms in more advanced societies. Yet the directions this change can take and its final products are constrained by the genetically influenced predispositions that constituted the earlier, simpler adaptations of preliterate human beings.[45]

It would appear that not only has Wilson forgotten his own characterisation of those 'basic social responses' to which he refers, but that the influence of the biological has shifted from that of an enabling mechanism, one rail of the 'dual track', to that of biological buffers on the lines of further cultural evolution. We are invited to contemplate the 'hard biological substructure' of 'elemental' categories of social behaviour which are inherited from hunter--gatherer existence and have 'substantially' influenced the course of subsequent cultural evolution.[46] The biological pole is clearly determinate in the last event. The details of the biological *deus ex machina* which Wilson introduces in order to extricate himself from the morass of contradictory arguments are not of immediate concern. As the critics of sociobiology have noted there is always some biological 'principle' by which the otherwise awkward or inexplicable fact may be safely slotted into a coherent picture of adaptive behaviours.[47] In this respect much of sociobiology when applied to human society becomes, in the words of Gould, merely 'speculative story telling'.[48]

What is more serious is the unconcealed biological reductionism which Wilson espouses in his explanations of adaptive behaviour. While, as Gould points out, clearly much of human behaviour can be seen as adaptive, the problem for sociobiology is that, in cultural evolution, humans have developed an alternative non-genetic system to support and transmit adaptive behaviour.[49] As Gould observes,

> An adaptive behaviour does not require genetic input and Darwinian selection for its origin and maintenance in humans; it

> may arise by trial and error in a few individuals who do not differ genetically from their groupmates in any way relevant to this behaviour spread by learning and imitation and stabilised across generations by value, custom and tradition.[50]

Sociobiology posits direct natural selection for specific behavioural traits, whereas in fact such behaviours may have more to do with structural determination than biological predispositions.[51] As Washburn suggests, while the genotype delimits a potential range of possible behaviours, the actual constraints so far as the transfer and accumulation of social learning is concerned are not primarily biological.[52] 'Knowledge and organisation are properties of groups and transmitted from generation to generation by learning, not genes.'[53] Indeed, Dobzhansky has argued, the human species is biologically such an extraordinary success precisely because its culture can change so much faster than its gene-pool. Cultural evolution has become adaptively 'the most potent extension' of biological evolution.[54] Human beings not only reproduce themselves organically and biologically, they reproduce themselves socially and they do this through specific forms of intervention in the natural world in the form of labour (the reproduction of their means of subsistence by the interposing of tools and language in the first instance). Their needs are not simply the expression of inherited 'predispositions', but, in the process of being met through labour activity, humans simultaneously create new needs, which in turn become a part of ongoing cultural development. The failure of sociobiology lies in its opposition to the notion of radical discontinuity between the social and the natural forms of development introduced by the emergence of human labour. While there is a dialectical relationship between the biological and human levels – and nowhere is this more obvious than in the emergence of early humans – nevertheless each level of development is characterised by particular regularities. As Dobzhansky has put it, warning of the dangers of reductionist explanations:

> the phenomena of the inorganic, organic, and human levels are subject to different laws peculiar to those levels. It is unnecessary to assume any intrinsic irreducibility of these laws, but unprofitable to describe the phenomena of any overlying level in terms of those of the underlying ones.[55]

This, as Dobzhansky himself acknowledges, is a basically Marxist perspective, but it arises on the ground of a scientific materialism which relies on concrete investigation of phenomena in the real world. One thing that the sociobiologists have consistently not done in this work on human behaviour is to test their theories by detailed research. Something that Wilson, in common with other sociobiologists has missed, as Sahlins points out, is the crucial function of speech for the human organism, not merely as a mode for carrying information, but crucially as a *structure of signification.*[56] It has already been suggested that what led the way over the 'highest implemental frontier' that separates the anthropoids from mankind was the intervention of some form of language in the process of tool-making. Like White, who had previously drawn attention to the importance of the symbolic faculty, Sahlins sees symbolisation as the key to the radical discontinuity between nature and culture which eventually freed human evolutionary development from the confines of biological determination.[57] In the final chapters, the question of the origin of human language is confronted.

6 Theories of language origins

Language leaves no trace in the fossil record. As such there is nothing in the way of direct evidence concerning the tantalising question of the origins of human language. Yet the emergence of linguistic symbolisation probably marked a decisive moment in the evolution of developing hominids towards a fully culture-bearing species. The recognition of the centrality of language in any explanation of the genesis of human culture has produced a wealth of speculative theories.[1] Here, in a brief review of some of the recent work on this issue, it is argued that the resolution of the difficulties of the various approaches described can only finally be achieved on the basis of a materialist explanation of language origins.

The obvious starting-point is the question left unanswered from the previous chapter, namely, the evolution of the human symbolic system out of the communication system of the primates. As Hockett, one of the pioneer theorists, has remarked,

> Until we can describe in detail just how human language differs from any variety of communicative behavior manifested by non-human or pre-human species, we cannot really know how much or little it means to assert this particular human uniqueness.[2]

Hockett itemises the basic 'design features' that mark out the distinctive features of human language as well as those shared in common with the communicative systems of other animals.[3] Of the thirteen different features discussed by Hockett only 'duality of patterning', whereby a relatively small stock of minimum meaningless phonemic ingredients can be ordered into a very large stock of morphemic combinations to which meanings can be assigned, is uniquely human.[4] The problem so far as the origin of speech is concerned, is how the 'closed' call system of the primates with a finite repertory of signals can be rendered 'open' and 'productive' among humans. The answer, it is suggested, lies in a 'blending' process whereby discrete elements of primate calls are fortuitously combined in a new combination to signal a novel situation. The example Hockett gives is the instantaneous blending of the call for food with the call for danger when the former is encountered at the same time as a predator is observed.[5] In itself this explanation is unsatisfactory since it poses a purely 'accidental' basis for the emergence of human speech through a mechanistic fusion of discrete elements. It is by no means clear in this view how perhaps the most crucial 'design feature' so far as the survival and adaptation of the species is concerned, that of 'displacement', might have come about.

Displacement may be defined as the ability to report on events not concurrent with the act of communication. Speech can be removed in time and place from the subject of discussion.[6] Displacement is, moreover, universal in human language. It enables experience to become cumulative and the lessons learned from earlier experience to be retrieved. In freeing communication from its immediate physical and temporal co-ordinates, this property of language opens up wholly new adaptive strategies based on the inherited transmission of cultural forms.[7] Interestingly, Hockett views transportation and shaping of tools for some future purpose as perhaps creating the circumstances for 'elements' of displacement to arise. However, the profound implications of this for the understanding of how language might have arisen are not properly grasped, and the scheme as a whole remains essentially eclectic and descriptive.

Greater theoretical consistency is to be found in the work of Hewes who has put forward a much acclaimed 'gestural' theory of language origins.[8] Hewes also addresses the question of the roots of human language in the communication of primates, pointing out that the primate call system is unlikely to have been the basis of human language being, 'mainly "emotional" and only meagerly propositional'.[9] Vocalisation in this view was not the 'initial pathway' to human communication. On the other hand, the work of the Gardners using manual signs with chimpanzees is cited by Hewes as possibly illuminating early forms of human communication, an observation which these writers themselves have taken up in their turn.[10] The central thrust of Hewes's argument is that speech would probably only have emerged with the hunting of large animals, where it would have facilitated the exchange of environmental information.[11] The handing down of tool traditions, says Hewes, probably depended for a long time not on speech but on visual observation in which the central element would be 'gestural imitation'.[12] Significantly, Hewes refers to Fouts's description of the 'moulding' process by which ASL was inculcated in Washoe.[13] Hewes suggests,

> the gesture or characteristic motor sequence associated with the making or use of a tool or weapon can come to serve as a sign for it ... The domain of tool and weapon names would arise in a gesture language from the distinctive patterns of movement implicated in their manufacture or use. Signs for operations such as pounding, cutting, pulling, twisting, throwing, smoothing or polishing, carrying, etc., are closely analogous.[14]

Hewes himself points out the disadvantages of a manual language in that it requires a fairly neutral background, good illumination, absence of intervening foliage, a relatively short distance between the transmitter and receiver and mutual frontal orientation. Making manual gestures is also slower, requires more energy and prevents the use of the hands for any other activity.[15] In view of this it is difficult to see why Hewes is so keen to ascribe to australopithecines a predominantly mimetic form of language. Although one can accept with Hewes that speech did not come into fully fledged existence *de novo,* the gestural theory still leaves unexplained the actual origin of spoken language. Indeed, if as Hewes suggests, following Lieberman, only post-Neanderthal man was physically capable of producing speech sounds and, moreover, the gestural system was itself effective for exchanging environmental information, it is difficult to imagine its subsequent replacement by a primarily vocal system.[16] Rather, it seems just as probable that such a gestural system would have been retained and refined in contemporary human communication. The fact is, however, that such gestural systems play only a secondary, at best qualifying, role in human communication. Despite the range and subtlety of non-verbal human communication, such forms are mainly auxiliary devices to spoken language.[17]

The degree to which gestures could be employed to transmit environmental information outside the immediate interactive context and across generations is also highly debatable. In other words, given the previous comments on displacement, it is open to question whether cultural traditions can be transmitted by purely gestural means and assimilated by observational learning. It should be remembered that these gestures are seen as *pre-vocal* in contrast to contemporary sign languages such as ASL, which are *derived* and interpreted from an existing speech basis. Hewes claims an upper limit of 2,000 gestures beyond which a gestural system would begin to yield to a vocal system of communication, a strictly quantitative criterion which implies a perspective on human language which locates it merely as a point on a continuum with that of other animals.[18] The actual transition to vocal sounds is posed as arising through their being 'systematically linked' with gestural language, whereby the vocal organs 'imitate' hand and other body movements, the so-called 'mouth–gesture hypothesis'.[19] Indeed, it could be argued that Hewes has added little to the understanding of the origination of human language beyond Darwin's claim 'that human language owes its origin to the imitation and modification, aided by signs and gestures, of various natural sounds, the voices of other animals, and man's own instinctive cries.'[20]

Although Hewes's work is useful in drawing attention to the interconnection between patterns of tool-activity and the emergence of language, his theory is deficient precisely because it fails to examine closely the social psychological process which would underly such a gestural language for it to become a viable cultural mode. As Hallowell has pointed out in a critique of Hewes,

> It is difficult, for example, to imagine how the manufacture of tools, and the development of tool making traditions could have arisen at a protocultural stage at which the mechanism of social transmission was exclusively observational learning and at which communication was mediated through signs rather than any form of symbolic representation.[21]

Hallowell suggests that the symbolic capacity, particularly in the form of speech, entirely altered the individual's relationship not only to his environment but to himself. The use of personal pronouns, personal names and kinship terms made it possible for the individual to symbolise and 'objectify' himself in a social framework.[22] Thus the individual could identify and reflect upon his conduct and that of others over time, within 'a common framework of socially recognised and sanctioned standards of behavior'.[23] Hallowell adopts the analytic framework of functionalist sociology, but the basis of his perspective is ego-psychology which is used in order to gain an insight into the 'psychological restructuralisation' leading to the development of the 'ego functions' of 'social actors' in early human society.[24] In so doing, all the assumptions of mid-twentieth century social science are projected onto the very dawn of human history, including such supposedly eternal features of human existence as 'property rights' over tools and all the subconscious burden of 'shame, guilt and anxiety' surrounding such institutions as incest taboo.[25]

Nevertheless, despite its idealist basis Hallowell has performed a useful service in drawing attention to a neglected dimension of evolution.[26] This is the complex issue of the developing social psychological capacities of the individual which will be re-examined in the final chapter. Hallowell, however, does not present an adequate basis for explaining the origins of human culture and language. Specifically, the origins of human language are accounted for only as aspects of enhanced 'self-awareness', an individualistic perspective, while the *social* roots of developing forms of consciousness are left almost entirely unexamined. This can be clearly seen in Hallowell's notion of 'protoculture', which is intended to emphasise the evolutionary continuities between primates and humans. The notion of protoculture is meant to provide a gradualist answer to the saltatory view that language and culture may have emerged 'suddenly' in the evolutionary process, or that there was some decisive 'cerebral Rubicon' as measured by increased brain size which separates off the human species. By contrast Hallowell argues that human culture is built upon a 'behavioral plateau' of 'preadaptive' primate behaviours.[27] Australopithecines, it appears, are protocultural creatures because their 'ego-centered processes remained undeveloped or rudimentary'.[28] The problem of how the transition from a protocultural to a cultural phase took place is simply left begging in this schema.

For Holloway also the central issue is of a social psychological nature, in this instance broadly 'how man organises his experience'.[29] In a critique of Hallowell's approach, Holloway attacks the consignment of australopithecines to the 'protocultural' stage on the spurious 'logical' grounds that some groups of developing hominids must have lacked the full evolutionary complement of 'design features' necessary for the attribution of human speech. [30] Holloway's own attempt, however, to establish 'the essential discontinuities' on a firmer palaeontological basis, using stone tools as evidence of 'the imposition of arbitrary form on the environment' by early humans, is equally fraught with difficulties.[31]

Specifically, Holloway sees chimpanzee tool-making using twigs as observed by Goodall and others, as 'iconic' activities in contrast to the 'non-iconic' nature of human tool-making. For humans, that is, there is no necessary relationship between the form of the original material and the shape of the finished tool.[32] Through the cultural capacity for symbolisation mankind is capable of imposing arbitrary structures on the environment.[33] It is not clear how the 'invention' of symbolisation first took place.[34] Holloway describes it merely as an 'emergent process' different in kind from anything which preceded it although linked to the emergence of 'standardised' tool-making techniques.[35] Several aspects of the models which describe language can, in Holloway's view, also be used to analyse tool-making.

> Both activities are concatenated, both have rigid rules about the serialisation of unit activities (the grammar, syntax), both are hierarchical systems of activity (as is any motor activity), both produce arbitrary configurations which, thence become part of the environment, either temporarily or permanently.[36]

In this context reference is made not only to the work of Hockett, but also to that of Saussure, Chomsky, Morris and Leroi-Gourhon, among others.[37] From these various writings Holloway argues that symbolisation can be seen as forms of 'social and material control' able to 'enforce' the arbitrary distinctions that became the basis for rule-governed and standardised tool-making procedures.[38] Such standardisation, imposed by rules analogous to grammatical rules, establishes the uniqueness of human culture. A grammar of tool-making is proposed paralleling the grammar of language and both are seen as deriving from common underlying cognitive structures.[39]

The major problem with Holloway's exposition is that it rests upon a reification of symbolic systems which betrays an essentially idealist bias. Whether or not intended, his characterisation of the power of symbolic forms comes close to linguistic determinism. The symbol, however, does not simply 'impose' form on the environment, but rather, insofar as that environment is the object of human concern and activity, it will be the collective summation of the social experience of labour which shapes and gives real content to the symbolic expression. The dialectical relationship between language and human activity therefore cannot be simply treated at the level of concordant 'cognitive structures' – important though the neurophysiological substrates undoubtedly are. It is confusion over this crucial issue which leads Holloway to assert he can see 'no good reason to claim that language must have followed tool-making', thereby assigning causal priority to the non-material factor.[40] In fact, Holloway's case remains unproven since in terms of the *emergence* of human symbolisation, stone tools represent probably a fairly late development in tool manufacture although the only evidence available in the palaeontological record.[41] So far as the relationship between tool-making and symbolisation is concerned, it is not iconicity which is important but, as previously suggested, the use of a tool to make another tool. Where this occurs, as the following chapter argues, there is evidence of a perceptual and cognitive restructuring of activity in a wholly new spatio-temporal dimension which may well imply the intervention of some form of language in the labour process. The character of symbolisation cannot be understood without the context of social activity in which it occurs. For developing humans this context was one in which socially co-operative labour was increasingly marking out the path of evolutionary advance towards a fully human status.

A further approach which attempts to link the emergence of language to labour is the 'environmental reference' theory of Lancaster. This also emphasises the discontinuities.[42] According to Lancaster, the ability to name objects provided humans with a way to refer to their environment and to communicate information about it to others. Lancaster maintains,

> An understanding of the emergence of human language rests upon a comprehension of factors that led to the evolution of a system of names. The ability to use names allows men to refer to the environment as opposed to the ability to express only his own motivational state. Object-naming is the simplest form of environmental reference. It is an ability that is unique to man.[43]

The ability to name objects would have had great adaptive value to a species evolving a new way of life different from other primates necessitating communication of information about the environment between members of the group.[44] Central to this new way of life is, says Lancaster, the utilisation of tools.

> Tool use in itself tends to focus attention on objects in the environment so that a stone or stick that has no significance or meaning in the life of a monkey, or at least none that need be communicated to others, may have great importance to the maker of primitive pebble tools.[45]

Thus raw materials or tools can be referred to and names given to plants and animals.

Like Holloway, Lancaster refers to the underlying neurophysiology of language, although in somewhat greater depth. The basis for understanding names is seen as lying in the reorganisation of the cerebral cortex which permits the formation of associations received in different sensory modalities, for example between sensations aroused in the visual cortex by the sight of the stimulus object and the sound name of the stimulus.[46] Such associations are independent of the limbic system, those older parts of the brain which all primates share, and which are primarily implicated in emotional behaviour.[47] Human language therefore can be used to refer to objects independently of emotion. A further feature of Lancaster's exposition which bears comment is the use of naturalistic studies of child language, not so much to represent a stage in the evolution of language, as 'a model for a simpler form of human language than that of adult humans'.[48] Lancaster notes that studies of two-word utterances of very young children contain mainly two classes of words. One is a very large class containing the names of objects and activities important to the life of the child, and the other is a much smaller class of 'pivot' or action words. Thus, language for the young child serves as an efficient system of environmental reference. It is in this sense that it is a 'model' for reconstructing the emergence of human language, rather than the now discredited theory of the ontogenetic development of the individual child's language somehow recapitulating the phylogenetic development of language in the entire human species.[49]

So far as the naming paradigm of language is concerned, however, certain important objections have been raised. In as much as it poses an arbitrary association between the word and the thing for which it stands, its referent, it is wholly compatible with a conditioning model of behaviour. As Limber points out, it follows therefore that any conditionable creature can learn human language and, indeed, this perspective underlies the various attempts to teach human language to chimpanzees previously reviewed.[50] However, an organism's use of 'names' is 'not sufficient evidence to conclude that the organism is using human language'.[51] The limitations this implies for the ability of the 'environmental reference' approach to go beyond a simple conditioning and essentially behaviourist approach can be seen in Lancaster's discussion of the relationship between tools and language. The link between these forms of activity remains purely external and essentially passive. The tool has an indicative function with respect to objects in the environment. However, the interconnection of labour and speech reshapes not simply the external environment, but the 'internal psychological environment' – the very structures of human psychic awareness. The emergence of *internally directed reference,* whereby the attempt to gain mastery over the external world results in successively greater control over the individual's own behaviour, must be included as part of the active dialectical relationship of labour and language. Lancaster's examination of the naming basis of language unfortunately does not account for this. It is not enough, moreover, to see the child's language as merely a 'simpler form' of adult language. Here too, there is a dialectical interrelation between the structure of the child's activity and its ability to form qualitatively *developing stages of generalisation* of the external world. The functioning of speech, as a means of 'verbal thought', in the words of the psychologist Vygotsky, is crucial here and is explored further in the concluding chapter.[52]

The emergence of human language can be compared to the emergence of language in the child, but only insofar as it is seen that the roots of each are in the social character of human activity. Vygotsky in fact shared Engels's view of the close interconnections of labour and language, describing the origins of human speech as, 'born of the need of intercourse during work'.[53] This perspective, which emphasised the social dimension, by no means has been confined solely to Marxist writers. In his pioneering work on language origins Noiré, who was contemporary with Engels, had observed,

> Language is the voice of the community. Even now the highest achievement possible is to order, direct and apply the forces of

> individuals to a common end, in united, organised activity, that is to say in work – for work is nothing but organised activity.[54]

For Noiré language stood in an 'indissoluble relation' to those human actions embodying various forms of labour to which the earliest meanings of verbal roots could be referred.[55] In this regard, it was proposed that the study of primitive stone tools and comparative philology could go hand in hand.[56]

In the first decades of the twentieth century, De Laguna's 'social control' theory was put forward in opposition to the prevailing psychologising approaches which resulted in 'metaphysical dualism' whereby speech was seen 'only as an external physical manifestation of inner psychical processes'.[57]

> What is needed is a fresh conception of speech as an essential activity of human life, fulfilling an indispensable function in the economy of life. It is in the light of this function that the causes and course of its development must be sought.[58]

For De Laguna also, speech was 'the great medium through which human co-operation is brought about' and its evolution could be seen as a necessary response to the need to organise and co-ordinate individuals for social labour.[59] Indeed, the very structure of language itself, the predicative nature which perhaps characterised mankind's first verbal utterances derived from this requirement.[60] Again, more recently, Révész in an important work, has argued in his 'contact theory' that early humans living in social groups would have been 'compelled to find suitable linguistic expressions for these vitally important activities' relating to the welfare of the group.[61] Echoing Noiré, Révész argues that the best clues to the emergence of human language probably lie in the study of stone tools, a cultural marker, and in particular, in the emergence of the making of 'standardised' tools which are held to presuppose language.[62]

> Man can only construct tools appropriate to a given end if he has the capacity to direct his activity toward it and is capable of surveying or visualizing the process of work to be carried out. Since conscious purpose and the invention of tools are based on mental processes that require linguistic fixation, we are justified in identifying *Homo faber* with *Homo loquens.*[63]

It could be argued, however, that Noiré, De Laguna and Révész in their seminal attempts to provide a materialist counter to idealist notions of language origins have perhaps erred in the direction of 'vulgar' materialism. The most satisfactory recent exploration of the issue of language origins and its connection with labour by a non-Marxist is to be found in the writings of the anthropologist Montagu.[64] Montagu attempts to integrate the logistical and cognitive aspects of language on the basis of a discussion of the Oldovai level I artifacts and fossil remains. Like Guilmet whose work was previously referred to, Montagu sees the early Oldowan tools as being not only skilfully made but as serving a variety of different usages. The making of such tools to a 'pre-existing pattern, calculated to serve a series of complex future purposes', including the making of other tools, is taken to imply a capacity for abstraction and the communication of such skills by verbal means.[65] These tools, it is suggested, were used mainly in the butchering of small animals.[66] Thus the view of Hewes and others, that it was only with the division of labour accompanying big-game hunting that the conditions arose for the development of speech, are rejected. Rather, Montagu claims that big-game hunting was an *effect* of the prior emergence of human speech. Although speech may have been at first rudimentary it had a profound effect both on labour itself and on the mental processes accompanying it.

> The growing mastery over the environment would have produced a novel form of interactivity between the members of the group in which speech facilitated the communication of ideas, practical ideas, relating to the manufacture of tools to serve various purposes in response to the challenges presented by the variety of animal food. In this way, in a quite complex feed-back interrelationship, small-game hunting, tool-making and speech would have reciprocally served to provoke each other to further advances. In this manner speech itself would develop as a special kind of tool designed to operate on man himself.[67]

Speech therefore provides the 'conceptual tools' or 'mental analogs' of material tools and Montagu puts forward the notion of conceptual and manual 'prehension' working together 'interoperatively'.[68] Developing hominids, at the australopithecine stage and perhaps even before, were thus able, says Montagu,

> to predict the utility of a hand-held tool beyond the period of

> immediate use, beyond immediately prehended relationships to more expansive ones in time and space. In short, with the domination of inert matter and the ability to mold it to one's needs for some future purpose must have gone the ability to conjure with the 'not-here' and the 'not-now'.[69]

The mediation of the interaction of hand, brain and tool through speech provides the materialist interpretation of the features of 'displacement' previously identified. The general capacity for 'reflection', the ability to plan and organise purposeful labour on the basis of the social accumulation of past labour probably therefore figured early in the evolution of humanity.

It has been suggested that the generalisation of previous labour experiences was perhaps performed at first through diffuse 'polysemantic' word notions in which were bound together all aspects of the labour process.[70] Only as the social life of developing hominids itself became more complex and diverse was the point reached at which words became the efficient 'tools' for the organisation of individual self-awareness, the conceptual analysis of the environment, the general communication of the aims and purposes of labour, as well as other subsidiary forms of social activity. The reconstruction of the transitional stages which led from polysemantic words to differentiated speech utterances is probably a fruitless exercise. On the other hand, the general character of human speech, in particular its interconnection with labour activity, is of importance. Thus, unsatisfactory and incomplete as theories of origins of language undoubtedly are, they do at least raise some central issues which require further examination. The final chapter attempts to bring to bear the insights of Marxist psychology onto the role of speech and tool-making in the development of human culture.

7 Labour and culture

The 'highest implemental frontier', the use of a tool to make a tool has been suggested as identifying the key discontinuity in the activity of hominids. Following Engels, it was argued that tool-making created the conditions for the historical inception of human speech. Both tool-making and speech represent indirect or mediated interventions in the natural and social world. Here the character of that mediation is examined from the standpoint of Marxist psychology.

Since the broad question of tool-making and speech has long been of interest to Marxist psychology, it is useful to begin by mentioning the seminal work of Pavlov upon which many more recent studies of this question have been built.[1] Pavlov's work on animal physiology eventually led him to regard human speech, which he termed 'the second signalling system', as a qualitative human attribute, different from anything possessed by animals.

> When the developing animal world reached the stage of man, an extremely important addition was made to the mechanisms of the nervous activity. In the animal, reality is signalised almost exclusively by stimulations and the traces they leave in the cerebral hemispheres ... This is the first system of signals of reality common to men and animals. But speech constitutes a second signalling system of reality which is peculiarly ours, being the signal of the first signals.[2]

These 'signals of signals' enable humans to link together multiple sensory stimuli and subsume them by unitary verbal signals or 'symbols' which are the key to conscious life and the human psyche. The connections arising on the basis of words introduce a new principle of neural activity, that of *generalisation* and *abstraction* from reality. Pavlov commented,

> For man, the word is just as real a conditioned stimulus as any other that he has in common with the animals, but at the same time the comprehensiveness of words is such that they cannot be compared either quantitatively or qualitatively with the conditioned stimuli of animals. As a result of an adult man's previous life experience, words are connected with all external and internal stimuli that reach the cerebral hemispheres; words signalize and stand for all these, and can therefore evoke all the actions and reactions in the organism that the stimuli themselves produce.[3]

This higher level of signalling interacts with and subordinates the lower, shared with the animals, and permits humans consciously to modify their behaviour.[4] Pavlov's achievement was to show that these generalised signals imparted a new quality into human orientation to the surrounding world – namely, a more profound reflection of reality whereby its essential features and relations could be distinguished and systematised.[5]

Language is socially acquired, as Pavlov's reference to the individual's 'previous life experience' implies, and to that extent words make available to the individual 'systems of connections impressed upon the brain as a result of earlier social activity'.[6] The work of Luria and Vygotsky has elaborated the importance of the social relationship with adults in the child's linguistic and conceptual development.[7] The child develops socially when, through speech, it masters other people's experience in a generalised form and, in principle at least, has open to it the entire conceptual advances and social experience of humanity. Subsequently, speech becomes the means whereby the child also organises its own experience and regulates and directs its own actions in a new way.

Mirroring the distinction drawn by Pavlov between first and second signalling systems, Luria distinguishes the impelling and significative functions of speech.[8] He shows how, in the course of the child's development, the character of the word as an impelling physical stimulus becomes subordinated to its symbolic functions, as it gradually becomes incorporated into the new system of verbal connections which alter the child's mental functions. By using speech for itself, internally, as it were, in the form of inner speech, the child can modify the relative natural strengths of the stimuli acting on it and is thus able to adapt its behaviour in an active and voluntary manner.[9]

Luria regards Pavlov's principle of the abstracting and generalising function of the second-signalling system as of decisive importance. Unlike animals, humans, having formulated a given rule as a generalisation, no longer need constant external reinforcement since words themselves provide this. 'A verbal formulation of a system of connections (in the shape of a rule which specifies each reaction) itself acts as a constantly operating reinforcement, sufficient to preserve the stability of a connection.'[10] It is the verbal-generalisation system which secures the formation and non-formation of new connections. Thus, in their orienting activity, humans have at their disposal a powerful, durable and flexible means of building up new functional systems on the basis of the formation of 'temporary speech connections'. It is this, says Luria, which gives human activity 'the character of "the highest self-regulating system" described by Pavlov'.[11] Pavlov and Luria provided the neurophysiological evidence for resolving the interrelationship of the biological and the social in language on the grounds of historical materialism. They have suggested how signs, particularly in the form of the word, enable human beings to master and subordinate the lower psychobiological functions and raise them to the level of cultural functions which identify the conscious and purposive character of human action.[12] The theoretical and experimental resolution of this question from a Marxist standpoint was first achieved by Luria's mentor and co-worker, Vygotsky.

Like Pavlov and Luria, Vygotsky understood the genesis and role of verbal signs as an integral part of the process of concept formation by means of which humans learn to organise and direct their actions.[13] Vygotsky's 'cultural-historical' approach sought to reveal the development of 'higher psychological functions' in the individual as specifically human, processes of 'mediation'. It should be noted that Vygotsky also began from Pavlovian neurophysiological premises.

> If, following Pavlov, we should compare the cerebral cortex to a 'magnificent signalling apparatus', then it might be said that man created the *key* to that apparatus – the magnificent signal system of speech – with whose help he masters the brain's activity. By means of speech he uses external stimuli to control his

> own behavior. No animal has anything remotely comparable with this.[14]

Signification, the creation and use of artificial signs was seen to introduce a new principle of activity which made possible man's active adaptation to nature. For Vygotsky, practical labour and theoretical cognitive activity were mediated through auxiliary means which are worked out socially and culturally transmitted. In practical activity mankind uses tools and in theoretical activity, signs.[15] While, however, the use of signs is analogous to the use of tools, in that both may be subsumed under the general notion of 'mediated activity', here the analogy ends. As Vygotsky shows, signs and tools possess wholly different orientations with respect to human behaviour.[16]

> The tool's function is to serve as the conductor of man's influence on the object of his activity; it is *externally* oriented; it must lead to change in objects. It is a means of man's external activity aimed at mastering, of triumphing over nature. The sign, on the other hand, changes nothing in the object of a psychological operation. It is a means of internal activity aimed at mastering man himself; the sign is *internally* oriented. These activities are so different from each other that the nature of the means they use cannot be the same in both cases.[17]

There is, however, a 'real tie' between these forms of activity in phylogenetic and ontogenetic development.

> The mastering of nature and the mastering of behavior are mutually linked, just as man's alteration of nature alters his own nature. In phylogenesis we can reconstruct this link through fragmented but convincing documentary evidence, while in ontogenesis we can trace it experimentally.[18]

As regards the latter question of the developmental history of the individual, in a series of important experimental observations Vygotsky and his colleagues traced the interrelation of tools and symbols in the course of the child's practical activity.They demonstrated how the incorporation of signs into the child's activity transforms that activity, penetrating the process of tool-use and producing 'fundamentally new forms of behavior'.[19] These new forms of behaviour moulded via tools and speech are the basis of the uniquely human active adaptation of nature to mankind's requirements.[20]

In his emphasis on the 'interweaving' of tools and speech Vygotsky followed upon and elaborated the ideas of Engels. In essence, Vygotsky revealed the dynamics of the 'higher psychological functions' as processes of reflecting reality. The theory of reflection was also present in embryo form in Engels's *Dialectics of Nature* but was, however, most fully worked out in philosophical terms by Lenin.[21]

In fact, Lenin's theory of reflection is regarded as having been given a materialist substantiation first by Pavlov's work on signalling systems, in particular, the generalising and abstracting function of the second-signalling system.[22] Although Pavlov was not himself a Marxist, he put forward a materialist and monist position which views consciousness as the highest organisation of matter.[23] The elementary form of reflection, common to all living organisms, is 'irritability' which at a higher stage of evolution becomes 'sensibility', the reflection of the properties of things in the form of sensation.[24] Human consciousness, however, is a form of reflection which is social by nature and inconceivable without the mastery of language. Lenin himself commented, 'Every word (speech) already *universalises* ... Language helps us make the transition from sense perception to generalised abstract thinking.'[25] The process of reflection is, however, not simply a passive one. The psychologist Lomov points out,

> Lenin's theory of reflection does not postulate a mere relation of adequacy of image (sensation, perception) to the reflected object. Higher forms of the process of reflection investigated by psychology are active processes. For a receptive image to ensue, the action of a single stimulus on the receptor is not enough, even though this action is an indispensable condition for the image to appear. Mental reflection is in its essence active, which means that the subject stands out in the process of cognition as an active agent.[26]

Integral to the process of reflection, therefore, is the activity of speech. Speech is new. For emerging hominids the increasing ability to assess the future consequences of their previous activity, through speech, would bestow important adaptive advantages over rival populations. Speech is also social. The reflection of objective reality which becomes essential for *planning future labour activity* is, so to speak, 'refracted through the prism of socially accumulated and generalised experience embodied in language.'[27] Thus mankind, through its labour activity, comes to acquire an understanding of the properties and laws governing the natural universe and is able to 'approximate' or adjust its behaviour in order, actively, to take account of the objective facets of the environment, including those changes brought about in it through labour activity.[28] Marx himself spoke of language as not simply the means of communication but as 'practical consciousness'.[29]

The early stage of the formation of consciousness Marx described as akin to an 'original awareness', that is, close to direct sensation of the environment – such consciousness being, as it were, 'at first, of course, merely consciousness concerning the *immediate* sensuous environment and consciousness of the limited connection with other persons and things outside the individual who is growing self-conscious'.[30] In pre-class society, insofar as humans distinguish themselves from nature and therefore from each other, theoretical and practical activity correspond. There is, albeit at a simple level, an 'identity' or coincidence of internal and external practical activity. Lukacs recognised this necessary elementary natural coincidence in his eventual acceptance of reflection theory, which he had repudiated in his earlier writings:

> the most primitive kind of work, such as the quarrying of stones by primeval man, implies a correct reflection of the reality he is concerned with. For no purposive activity can be carried out in the absence of an image, however crude, of the practical reality involved. Practice can only be a fulfilment and a criterion of theory when it is based on what is held to be a correct reflection of reality.[31]

Human activity presupposes the optimisation or 'efficient' use of time and energy in the satisfaction of needs.[32] Time is the condition of activity. As Marx noted in the *Grundrisse,*

> Economy of time, to this all economy ultimately reduces itself. Society likewise has to distribute its time in a purposeful way, in order to achieve a production adequate to its overall needs ... economy of time, along with the planned distribution of labour time among the various branches of production, remains the first

> economic law on the basis of communal production. It becomes law, there, to an even higher degree.[33]

While it was suggested that the overall distribution of labour time among early humans can only be guessed at, it is clear that their activity already took place within a decisively altered temporal context. The use of tools to manufacture tools implies activity carried out with some 'ideal' mental image of the future results of labour already in mind.[34] Dobzhansky has provided a formulation of this point.

> The capacity to form and to operate with abstract ideas is correlated in evolution, if not in physiology, with the capacity to use human language. Here too, the product grows with the instrument and vice versa. And these capacities are, in turn, correlated with toolmaking. It should be noted that tool using and toolmaking are performances as profoundly different as signs and symbols. To make a tool for a future employment one needs more than manual dexterity; what is necessary is formation of a mental picture of a situation which is expected to arise in the future but which is not yet given to the senses.[35]

Herein, then, lies the function of what to begin with must have been the most rudimentary forms of symbolic speech, namely, its 'time-binding' property which was earlier described as 'displacement'. Critchley, following Pavlov, has described this as follows,

> Man's utterances entail the use of symbols or signs of signs and consequently possess the superlative advantage of applying to events in time past, present, and future and to objects *in absentia*. This endowment has been called the 'time-binding' property of human language. It also possesses the merit of beginning the storage of experience.[36]

Speech, then, brought with it 'release from the shackles of time present', and for evolving hominids the entry into a new temporal dimension of existence increasingly governed by the 'economy of time'.[37] The objective basis of this economy lay in an increasing dependence on the 'purposeful' manufacture of tools with which to perform labour, including the fashioning of other tools, rather than on spontaneous and unmediated forms of expenditure of energy. Tools also have a 'time-binding' property. Tools not only mediate between the activity of the producer and the object of labour but also the general social relations within which activity takes place. Tools are as much the products of labour as they are its instruments. They, too, are 'objectivisations' of the accumulated labour experience of humanity. In this regard, tools are not just instruments of immediate activity but embodiments of previous activity as well as projects of future activity. The economy of time, then, becomes identical with the inauguration and development of artificial productive forces.[38] It is this inauguration of artificial productive forces, doubtless initially extremely tentative, but interpenetrating ever more closely with time-binding symbolic speech, which sets in motion the economy of time that lies at the base of human culture.

White has defined culture as 'a mechanism for harnessing energy'.[39] In choosing between various assignments of energy the very first social estimation of economy of labour time would have emerged in the primitive forms of the division of labour. Sahlins has suggested, however, that a *surplus* of disposable time exists after subsistence needs have been met in primitive hunter–gatherer forms of society, which are characterised as being 'the original affluent society'.[40] In opposition to White, Sahlins argues that in terms of energy harnessed *per capita*, neolithic economies were no more productive than palaeolithic even though the total social product of the former is greater, due to the larger population which it supports. Sahlins, however, fails to appreciate the essential point that, as compared to hunter–gatherer society, farming and domestication of animals represent definite advances in mankind's understanding of the laws of nature. As such, this represents a potentially *more efficient* utilisation of tools and productive forces, and the greater planning of labour activity. As Engels noted, 'All our mastery of it [nature] consists in the fact that we have the advantage over all other creatures of being able to learn its laws and apply them correctly.'[41] Sahlins's notion, therefore, runs counter to that of 'economy of time' and the proper proportional distribution of labour which becomes a necessity even in the earliest forms of society.[42]

The significance of time as a factor in human evolution has been explored by Pavlovian physiologists who have examined the structure of the central nervous system and the nature of the conditional reflex and, in demonstrating the principle of 'anticipatory reflection', have attempted to reveal the physiological character of labour activity as a process which *anticipates the future*.[43] A broader neuropsychological reconstruction of the basic relations between the labour process and the brain structures of hominids at various stages of development has been proposed by Kochetkova as part of her monumental work on palaeoneurology.[44] On the basis of the study of fossil endocasts, Kochetkova identified the unequal development of individual areas of the cerebral cortex. Specific 'intensive growth foci' are seen as localised in particular sub-regions of the cortex which 'take up a considerable area in modern humans', as compared say to anthropoids or early hominids.[45] Those areas, it is suggested, are implicated in more complex tool-making and speech, and are correlated with detailed evidence from endocasts of the hominid forms which succeeded the australopithecine stage. It can be argued that this evidence points towards the formation of new systems of neural organisation, permitting the successive stabilisation of progressively human adaptation through labour activity and its associated 'higher psychological functions'.

It may be, however, that the neuropsychological preconditions for the emergence of these new systems of cortical connections were laid down and already present, to some extent, in the australopithecines. It has been noted, by Campbell, for instance, that the roots of language 'in the form of new cortico-cortical connections may have appeared well before the brain expanded'.[46] The increase in brain size, therefore, can be viewed as the flowering of the process of human adaptation and not simply its inception.[47] It is possible that somewhere in the period described earlier, as marked by a 'fossil void', between the ramapithecines and australopithecines, there occurred the initial transformation of neuropsychological structures that first permitted developing hominids to advance from merely tool-using creatures to tool-dependent tool-makers who had crossed the 'highest implemental frontier', and who employed some rudimentary linguistic means in their labour activity. Such a transformation would have constituted a 'quantum leap' in terms of the new adaptive possibilities which were opened up. Engels himself, indicated the existence of such 'nodal points' which he held to 'determine the various qualitative modes of matter in general'.[48] Subsequently, this new potentiality may have required millions of years to unfold before the eventual extrication of human beings from purely biological determinates became manifest in the fossil record as the beginnings of a recognisably developing cultural mode of adaptation.

In the same way that the roots of human tool-making can, as Engels suggested, be traced back prior to the subsequent development of the human hand, so also, then, the foundations of human culture can be traced back to a phase of evolution which preceded the expansion of the human brain. Much of what can be thought of as peculiarly human in terms of the organisation of the human brain is, therefore, more properly considered as a result of cultural adaptation than its cause. As Washburn has put it: 'From the immediate point of view, this brain makes culture possible. But from the long-term evolutionary point of view, it is culture which creates the human brain.'[49] Indeed, it is the interconnection of hand and brain in labour activity which constitutes the real framework of culture and is eventually reflected in the unique, specifically human, shape of the cortex. The brain, therefore, can be regarded not only in terms of its function as an 'organ of labour' but, like the developing hand with which it has intimate neuro-sensory connections, also as the necessary *product* of labour. This evolutionary interconnection between tool, hand and brain was noted by Engels.[50] It is given modern formulation by Washburn and Lancaster.

> It appears that the form of the human hand, the large area of the brain directly related to the hand, the much larger areas of the cortex related to skilfull motor activity, and the greatly expanded cerebellum, also related to skilfull activity, all evolved long after initial tool use and in response to new selection pressures arising from the success of implements of many kinds.[51]

Those areas of the cortex which are expanded can be specifically related to the general requirements for successful labour, in particular the large areas concerned with the hand, speech, increased memory and planning.[52] The whole 'selective milieu' of the species is thereby altered. As Leontiev observes,

> The experience of the species is now reflected not in changing, let us say, the human hand, but in changing the implement used by this hand; it is reflected in ways and modes of using this implement which are fixed and generalised in it.[53]

An 'inverse relation' emerges between the transformation of the physical form of mankind and the modification of the form of labour. For early hominids insignificant modifications in technique were accompanied by substantial changes in morphology. In *Homo sapiens* major transformations in technique and forms of activity have little importance for their physical form.[54] The terms of this 'inverse relation' are strictly materialistic but can easily give rise to an idealistic view of the brain as the primary factor in human development, while, as Engels put it, 'the more modest productions of the working hand retreated into the background'.[55] It was precisely on these grounds that Engels originally formulated his essay on human origins, and chastised his contemporaries for their idealism.

> All merit for the swift advance of civilisation was ascribed to the mind, to the development and activity of the brain. Men became accustomed to explain their actions from their thoughts, instead of their needs – (which in any case are reflected and come to consciousness in the mind) – and so there arose in the course of time that idealistic outlook on the world which, especially since the decline of the ancient world, has dominated men's minds. It still rules them to such a degree that even the most materialistic natural scientists of the Darwinian school are still unable to form any clear idea of the origin of man, because under this ideological influence they do not recognise the part that has been played therein by labour.[56]

The anthropologist Leacock has given an admirable and succinct formulation of the kernel of Engels's central hypothesis in 'The Part Played by Labour in the Transition from Ape to Man'. It sums up the whole substance of the Marxist approach to labour, language and human culture.

> It was through labor that humanity created itself as a skilfull, large-brained, language-using animal, and through labor that it created an elaborate cultural superstructure. The very impressiveness of mankind's mental achievements, however, has obscured the fundamental significance of labor. Furthermore, the separation of planning for labor from the labor itself, a development of complex society, contributed to the rise of an idealistic world outlook, one that explains people's actions 'as arising out of thoughts instead of their needs'.[57]

The preceding argument has attempted to suggest that the 'fundamental significance of labor' can only be fully appreciated if the uniquely interrelated developments of tool-making and speech are understood as the broad path of evolutionary transition from anthropoid to humanity. It is the constitution of human activity as social labour which provides the materialist basis for human development to take place within the framework of culture and history.

Conclusion

This book began by outlining Engels's main hypotheses concerning the evolution of humanity. Recent archaeological finds, in particular East African materials, were examined in the light of these. The question of hunting and gathering as the earliest form of social labour was examined and some suggestions from contemporary anthropology were offered as to the role of these activities in human development. This issue was again taken up in an examination of the possible light that primates might shed on human origins and the position put forward was that a very special kind of tool-making characterises human labour activity. It was also argued that the role of speech in the development of cultural traditions was unique to human beings. The contention that primates also possess cultural traditions and linguistic capabilities was examined and a number of criticisms were advanced of those writers who attempt to obscure important discontinuities. Finally, the controversial issue of language origins was taken up and, after reviewing some of the existing theories, an approach based on Engels and subsequently developed by Marxist psychology was outlined, linking together tool-making and speech with the notion of reflection. This is described here as the labour theory of culture to differentiate it from idealist approaches to the notion of human culture.

In conclusion, it should be said that the argument that has been presented here, like Engels's original essay, is incomplete in several respects. The evidence which is available concerning the decisive phases of transition towards humankind is not sufficient to allow the claim that the interpretation that has been advanced is in any way definitive. Moreover, the framework of dialectical materialism within which this interpretation has been made provides no ready-made formula for the solution of any given problem, particularly a problem of such complexity as the origins of humanity. In itself, it offers only a general guide to what may be seen as significant questions relating to the development of social and natural phenomena in their changes, continuities and discontinuities. For Marxism, but not just Marxism alone, one such significant question is the development of human beings as a cultural species and the part played in this by labour.

Appendix

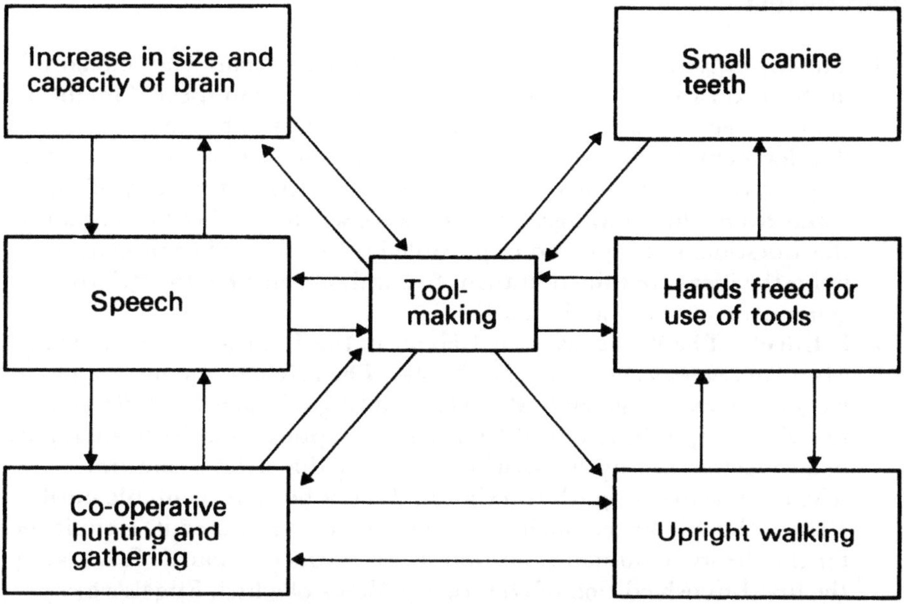

Diagram of the interconnected process of hominisation through labour (adapted from R. E. Leakey and R. Lewin, *Origins*)

Notes

Introduction

1 The writer is aware that the use of such terms as 'man' and 'mankind' in this text rather than 'human being' or the 'human species' might be held to suggest that only one sex played any part in human development. It is hoped that the arguments which are presented will dispel any fears of such a distortion. Nevertheless, since many of the works from which passages are cited were written without attention to this question, it has proved impossible in the course of commenting upon the ideas contained in them to eliminate entirely the strictly generic use of 'man' and 'mankind'.

2 F. Engels, 'The Part Played by Labour in the Transition from Ape to Man', in *Dialectics of Nature,* pp. 279-96. The original manuscript of Engels's essay was given to Bernstein on Engels's death in 1895 and remained unpublished until 1927, when it appeared in Russian as part of a collection of Engels's writings applying dialectics to natural science. The first English translation did not become available until 1940, although writers such as Bernal had already noted its significance for the theory of human evolution several years previously. Reviewing the first English edition of *Dialectics of Nature* of which Engels's essay comprises chapter 9, Bernal commented,

> This is probably the most important part of the whole book because in it we find most definitely stated the view that man owes his uniqueness to the existence of society and that society in the first place is common labour.

See J. D. Bernal, *The Freedom of Necessity,* p. 364.

3 See E. Reed, 'Evolutionism and Anti-Evolutionism' (1957), in *Sexism & Science,* pp. 161-77. See also B. G. Trigger, 'Engels on the Part Played by Labor in the Transition from Ape to Man: an Anticipation of Contemporary Anthropological Theory', *Canadian Review of Sociology and Anthropology,* vol. 4, no. 3, 1966, pp. 165-76. Trigger's essay provides a valuable review of Soviet work in this field. A more recent review of relevant Soviet and East European studies is contained in L. S. Klejn, 'A Panorama of Theoretical Archaeology', *Current Anthropology,* vol. 18,

no. 1, 1977, pp. 1-42 and vol. 18, no. 2, 1977, pp. 371-3. A recent Soviet appreciation of Engels's contribution to the study of 'anthropogenesis' is I. Andreyev, 'The Origins of Man and Society', *Social Sciences,* USSR, vol. 8, no. 3, 1977, pp. 101-12. The work of Engels is also discussed fully in G. Patšch, *Grundfragen Der Sprachtheorie,* especially pp. 67-82.

A critique of Engels's essay by a Western non-Marxist whose own work is otherwise informed by a sympathetic appreciation of Soviet scholarship is F. E. X. Dance, 'Speech Communication in the Soviet Union: the Phylogenesis of Speech According to Frederick Engels', *The Speech Teacher, Journal of Speech Association of America,* vol. 13, 1964, pp. 115-18.

4 See for example S. Diamond (ed.), *Towards a Marxist Anthropology: Problems and Perspectives.* Engels's work receives only brief dismissive reference. For general attempts to downgrade Engels's contribution to the development of Marxist thought see N. Levine, *The Tragic Deception: Marx contra Engels.*

5 J. Lewis, *The Uniqueness of Man.*

6 See J. Hoffman, 'The Dialectics of Nature: "the Natural–Historical Foundation of our Outlook" ', *Marxism Today,* vol. 21, no. 1, 1977, pp. 11-18. This view of the identity of outlook between Marx and Engels on the question of the dialectic in nature has been sharply contested, most recently by T. Carver, 'Marx, Engels and Dialectics', *Political Studies,* vol. 28, no. 3, 1980, pp. 353-63.

7 For a general review of this and related issues see K. M. Zavadskii, A. M. Georgievskii and A. P. Mozelov, 'Engels and Darwinism', *Soviet Studies in Philosophy,* vol. 10, no. 1, 1971, pp. 63-80. See also T. Ball, 'Marx and Darwin: a Reconsideration', *Political Theory,* vol. 7, no. 4, 1979, pp. 469-84.

8 C. Darwin, *The Descent of Man.*

1 Engels and human origins

1 K. Marx, *Economic and Philosophical Manuscripts of 1844,* pp. 112-14.

2 K. Marx, *Capital,* vol. 1, p. 177.

3 F. Engels, 'The Part Played by Labour in the Transition from Ape to Man', in *Dialectics of Nature,* p. 279.

4 Ibid., p. 281 (original emphasis).

5 Ibid., pp. 17-18 (original emphasis).

6 Cf. Marx, *Economic and Philosophical Manuscripts of 1844,* p. 113.

7 Marx, *Capital,* vol. 1, p. 179

8 Engels, op. cit., pp. 282-3 (original emphasis).

9 Ibid., p. 284. The problem of whether labour can *ever* exist without speech and thought and, if not, how it can be said that labour precedes articulate speech, albeit in the first instance, has been described by Hoffman as one of 'priority within unity'. See J. Hoffman, *Marxism and the Theory of Praxis.* Such a 'priority within unity' may be said to exist 'where one is both able to locate an ultimately determining factor

which nevertheless only exists in *interaction* with all the rest.' Thus, writes Hoffman, 'while strictly speaking, human labour has to be accompanied by speech and thought to be *properly* human, the corporeal activity which *became* human labour preceded speech and thought, hence establishing its causal priority as the *material* factor'. J. Hoffman (personal communication).

10 Engels, op. cit., p. 285.

11 Ibid., pp. 288-9.

12 Ibid., p. 291. Engels goes on to add a cautionary note which has a remarkably contemporary ring to it: 'Let us not, however, flatter ourselves overmuch on account of our human conquest over nature. For each such conquest nature takes its revenge on us' (pp. 291-2).

13 Ibid., pp. 295-6.

14 C. Darwin, *The Descent of Man*, p. 95 and p. 104. For discussion of revisions to the Darwinian picture of gradual evolution and its implications for human development see below chapter 7, note 48.

15 For a vigorous attack on the basic thesis that man produces himself through labour, on 'ontological' grounds see J. F. Crosby, 'Evolutionism and the Ontology of the Human Person: Critique of the Marxist Theory of the Emergence of Man', *The Review of Politics*, vol. 38, no. 2, 1976, pp. 208-43. Crosby, however, does not reject the notion of man's uniqueness but in an attempt to 'vulgarise' Engels, he suggests that perhaps Engels does.

16 Marx, *Capital*, vol. 1, p. 372, note 3.

17 Ibid.

2 The fossil record

1 G. E. Lewis, 'Preliminary Notice of New Man-Like Apes from India', *American Journal of Science* (5th Series), vol. 27, no. 159, 1934, pp. 161-82.

2 E. L. Simons, 'The Early Relatives of Man', *Scientific American*, vol. 211, no. 1, 1964, p. 62. To date there are unfortunately no post-cranial remains of *Ramapithecus.*

3 S. L. Washburn, 'Tools and Human Evolution', *Scientific American*, vol. 203, no. 3, 1960, p. 69. For an opposing view see W. G. Kinsey, 'Evolution of the Human Canine Tooth', *American Anthropologist*, vol. 73, no. 3, 1971, pp. 680-94.

4 E. L. Simons, *'Ramapithecus'*, *Scientific American*, vol. 236, no. 5, 1977, p. 33. See also P. Andrews, 'Hominoid Habitats of the Miocene', *Nature*, vol. 289, 1981, p. 749. Andrews argues that the evidence of habitat now points to ramapithecines as being omnivorous.

5 J. Napier, 'The Antiquity of Human Walking', *Scientific American*, vol. 216, no. 4, 1967, p. 64. See also J. D. Clark, 'African Origins of Man the Tool-Maker', in G. L. Isaac and E. R. McCown (eds), *Human Origins*, p. 6.

6 C. F. Hockett and R. Ascher, 'The Human Revolution', in M. F. A. Montagu (ed.), *Culture: Man's Adaptive Dimension*, pp. 20-49.

7 Ibid., pp. 29-30.
8 Ibid., p. 30.
9 D. S. Coursey, 'Hominid Evolution and Hypogeous Plant Foods', *Man* (ns), vol. 8, no. 4, 1973, pp. 364-5. Comment on A. Mann, 'Hominid and Cultural Origins', *Man* (ns), vol. 7, no. 3, 1972, pp. 379-96.
10 C. J. Jolly, 'The Seed-Eaters: a New Model of Hominid Differentiation Based on a Baboon Analogy', *Man* (ns), vol. 5, no. 1, 1970, pp. 5-26.
11 Coursey, op. cit., p. 365.
12 C. L. Brace and A. Montagu, *Human Evolution*, p. 282.
13 L. S. B. Leakey, 'Adventures in the Search for Man', *National Geographic*, vol. 123, January 1963, pp. 132-52. Leakey's designation of the finding of a rough cobble as a stone tool or 'manuport' along with the Fort Ternan remains is not generally accepted.
14 Simons, 'The Early Relatives of Man', p. 62.
15 See R. Leakey and R. Lewin, *People of the Lake*, pp. 27-8. Taxonomic problems remain, however, with at least two other probably related genera *Sivapithecus* and *Gigantopithecus* showing some hominid characteristics, such as enamelling, in their dentition. See Simons, '*Ramapithecus*', p. 28. Reports of several of the finds are listed as follows: P. Andrews and H. Tobein, 'New Miocene Locality in Turkey with Evidence on the Origin of *Ramapithecus* and *Sivapithecus*', *Nature*, vol. 268, 1977, pp. 699-701; D. Philbeam, G. F. Meyer, C. Badgley, M. D. Rose, M. K. L. Pickford, A. K. Behrensmeyer and S. M. I. Shah, 'New Hominoid Primates from Siwaliks of Pakistan and their Bearing on Hominoid Evolution', *Nature*, vol. 270, 1977, pp. 689-95; M. Kretzoi, 'New Ramapithecines and *Pliopithecus* from the Lower Pliocene of Rudábanya in North Eastern Hungary', *Nature*, vol. 257, 1975, pp. 578-81.
16 See R. Leakey and R. Lewin, *People of the Lake*, pp. 39 ff. Simons '*Ramapithecus*', p. 33, suggests *Ramapithecus* may have given rise to *Australopithecus*. Unlike *Ramapithecus*, no *Australopithecus* specimens have been found as yet outside Africa. Molecular biological analyses of blood serum proteins of humans and chimpanzees have been used to suggest that the evolutionary divergence of each is as recent as 5 million years B.P. If so, this rules out *Ramapithecus* as a direct ancestor of humans. See V. M. Sarich and A. C. Wilson, 'Immunological Time Scale for Hominid Evolution', *Science*, vol. 158, 1967, pp. 1200-03. Other calculations, however, place the time of divergence nearer 14 million years ago. See C. O. Lovejoy, A. H. Burstein and K. G. Heiple, 'Primate Phylogeny and Immunological Distance', *Science*, vol. 176, 1972, pp. 803-5.
17 L. S. B. Leakey, 'Finding the Earliest Man', *National Geographic*, vol. 118, September 1960, pp. 420-35. See also L. S. B. Leakey, 'A New Fossil Skull from Olduvai', *Nature*, vol. 184, 1959, pp. 491-3.
18 Considerable controversy has arisen as to whether or not *Australopithecus* or *Homo* was a tool-*maker* as against simply a tool-*user*. The Leakeys have ascribed tool-making to the even larger hominid *Homo habilis* whose remains were recovered at Olduvai shortly after '*Zinjanthropus*'.

This was estimated to have a cranial capacity of 680cm^3 and was therefore held to be a 'true man'. See M. Leakey, *Olduvai Gorge,* vol. 3, p. 49. The attempt to reserve tool-making to larger-brained species which are classified as *Homo* follows the tradition of speculative idealism which asserts the primacy of brain size as the criterion of humanness. For an opposing view to the Leakeys' on the question of australopithecine tool-making see M. H. Wolpoff, 'Analogies and Interpretation in Palaeoanthropology', in C. Jolly (ed.), *Early Hominids of Africa,* pp. 461-502.

19 See Jolly, ibid. For a summary of these researches, see R. E. Leakey and R. Lewin, *Origins,* especially pp. 78-117.

20 S. R. Binford and L. R. Binford, 'Stone Tools and Human Behavior', *Scientific American,* vol. 220, no. 4, 1969, pp. 70-84.

21 Ibid., p. 71.

22 G. L. Isaac, 'The Food-Sharing Behavior of Protohuman Hominids', *Scientific American,* vol. 238, no. 4, 1978, pp. 90-108.

23 Ibid., p. 96.

24 D. N. Stiles, R. L. Hay and J. R. O'Neil, 'The MNK Chert Factory Site, Olduvai Gorge, Tanzania', *World Archaeology,* vol. 5, no. 3, 1974, pp. 285-308. The initials here and in subsequent citations are for the museum reference and/or site discoverer.

25 Ibid., pp. 292-3.

26 Ibid., p. 291.

27 P. R. Jones, 'Effect of Raw Materials on Biface Manufacture', *Science,* vol. 204, 1979, pp. 935-6. Bifaces begin to appear at Olduvai in the middle Bed II dating from about 1.5 million years ago. See M. Leakey, *Olduvai Gorge,* p. 269.

28 P. R. Jones, 'Experimental Butchery with Modern Stone Tools and its Relevance for Paleolithic Archaeology', *World Archaeology,* vol. 12, no. 2, 1980, pp. 152-65.

29 The pioneering microwear study was conducted by the Soviet archaeologist Semenov. See S. A. Semenov, *Prehistoric Technology,* first published in Russian in 1957. An up-to-date review of this highly technical field is published in B. Hayden (ed.) *Lithic Use-Wear Analysis.*

30 See D. C. Johanson, M. Taieb, Y. Coppens, and H. Roche, 'Expédition internationale de l'Afar, Ethiopie (4e et 5e campagnes 1975-7), Nouvelles découvertes d'hominidés et découvertes d'industries lithiques pliocènes à Hadar', *Comptes Rendus de l'Académie des Sciences* (Series D.), vol. 287, 1978, pp. 237-40. For other tool finds associated with fossil hominids, see G. L. Isaac, 'The Archaeological Evidence for the Activities of Early African Hominids', in Jolly, *Early Hominids of Africa,* pp. 219-54. See also G. L. Isaac, 'Plio-pleistocene Artifact Assemblages from East Rudolf, Kenya', in Y. Coppens, F. C. Howell, G. L. Isaac and R. E. F. Leakey (eds), *Earliest Man and Environments in the Lake Rudolf Basin,* pp. 552-64.

31 R. E. F. Leakey, 'Evidence for an Advanced Plio-pleistocene Hominid from East Rudolf, Kenya', *Nature,* vol. 242, 1973, pp. 447-50 and R. E. F. Leakey, 'Skull 1470', *National Geographic,* vol. 143, June, 1973, pp. 818-29.

32 A. Walker and R. E. F. Leakey, 'The Hominids of East Turkana', *Scientific American*, vol. 239, no. 2, 1978, pp. 44-56. The exact age of the KBS-tuff from below which KNM-ER 1470 comes is disputed. Potassium/Argon analysis originally dated it at 2.61 (±·26) million years old. New calculations have revealed much younger dates between 1.82 (±·04) million years and 1.60 (±·45) million years, giving a difference between the oldest and youngest proposed KBS-tuff dates of 1.3 million years.

33 D. C. Johanson and M. Taieb, 'Plio-pleistocene Hominid Discoveries in Hadar, Ethiopia', *Nature*, vol. 260, 1976, pp. 293-7. Issues of phylogenetic relationships are notoriously contentious and comprise a voluminous literature. For a readable introduction, see Brace and Montagu, *Human Evolution*. Walker and Leakey, 'The Hominids of East Turkana', suggest five possible classification systems. See also Clark (in Isaac and McCown, *Human Origins*, p. 15) who suggests the possibility that the 'robust' and 'gracile' forms of *Australopithecus* may even represent the extremeties of sexual dimorphism within a single lineage. J. Reader, *Missing Links*, pp. 213-26, records the heated dispute between R. Leakey and Johanson over the classification of African material. See D. C. Johanson and T. D. White, 'A Systematic Assessment of Early African Hominids', *Science*, vol. 203, 1979, pp. 321-30.

34 See D. C. Johanson and M. Edey, *Lucy, the Beginnings of Humankind*.

35 M. D. Leakey and R. L. Hay, 'Pliocene Footprints in the Laetolil Beds at Laetoli, Northern Tanzania', *Nature*, vol. 278, 1979, p. 323.

36 Ibid., p. 320.

37 J. Napier, 'The Antiquity of Human Walking', pp. 55-66.

38 See C. O. Lovejoy, G. H. Heiple and A. H. Burstein, 'The Gait of *Australopithecus*', *American Journal of Physical Anthropology*, vol. 38, no. 3, 1973, pp. 757-80.

39 J. Napier, 'The Evolution of the Hand', *Scientific American*, vol. 207, no. 6, 1962, p. 62.

40 M. Day, 'Hominid Postcranial Material from Bed I, Olduvai Gorge', in Isaac and McCown, *Human Origins*, p. 368.

41 See Napier, 'The Evolution of the Hand', p. 62.

42 Ibid., p. 62. See also R. L. Susman and N. Creel, 'Functional and Morphological Affinities of the Subadult Hand (OH7) from Olduvai Gorge', *American Journal of Physical Anthropology*, vol. 51, no. 3, 1979, pp. 311-32.

43 R. L. Holloway argues on the basis of endocranial casts of early hominids that encephalisation is not a significant feature, the size of the brain having remained proportionally similar relative to body weight since the australopithecines. See R. L. Holloway, 'The Casts of Fossil Hominid Brains', *Scientific American*, vol. 231, no. 1, 1974, pp. 106-15. For an opposing view to Holloway see H. J. Jerison, 'Discussion Paper: the Paleoneurology of Language', in S. R. Harnad, H. D. Steklis and J. Lancaster (eds), *Origins and Evolution of Language and Speech*, New York Academy of Sciences, vol. 280, 1976, pp. 370-82. See also discussion in chapter 7 below.

44 C. Geertz, 'The Transition to Humanity', in S. Tax (ed.), *Horizons of Anthropology*, p. 42.
45 S. L. Washburn, 'Tools and Human Evolution', p. 69. See also Darwin, *The Descent of Man*, pp. 140-8.

3 Hunting and gathering

1 J. B. Lancaster, 'On the Evolution of Tool-Using Behavior', *American Anthropologist*, vol. 70, no. 1, 1968, p. 57.
2 S. L. Washburn and I. De Vore, 'Social Behavior of Baboons and Early Man', in S. L. Washburn (ed.), *Social Life of Early Man*, p. 94.
3 G. L. Isaac, 'The Food-Sharing Behavior of Protohuman Hominids', *Scientific American*, vol. 238, no. 4, 1978, p. 106.
4 Washburn and De Vore, op. cit., p. 97. Loss of body hair may also have been a factor here depriving the infant of an additional means of clinging to the mother.
5 Ibid., p. 98. Washburn and De Vore suggest loss of oestrus would have accompanied continuing child-care.
6 A. Mann, 'Hominid and Cultural Origins', *Man* (ns), vol. 7, no. 3, 1972, p. 382.
7 Ibid., p. 384.
8 M. Leakey, *Olduvai Gorge*, vol. 3, p. 260.
9 Ibid., p. 261. The DK site is dated at about 1.9 million years old.
10 L. B. Freeman, 'A Theoretical Framework for Interpreting Archaeological Materials' in R. B. Lee and I. De Vore (eds), *Man the Hunter*, pp. 264-5. See also S. R. Binford, 'Ethnographic Data and Understanding the Pleistocene', in ibid., p. 274.
11 R. B. Lee, *The !Kung San*, p. 433.
12 Ibid., pp. 116-57.
13 Ibid., p. 117.
14 Ibid., p. 460. See also L. A. White, *The Evolution of Culture*, pp. 256-60, and E. B. Leacock, introduction to F. Engels, *The Origin of the Family, Private Property and the State*, pp. 7-67 for further discussions of primitive communism.
15 Lee, *The !Kung San*, p. 244.
16 Ibid., p. 249.
17 Ibid., p. 246.
18 Ibid., pp. 149-51.
19 Ibid., p. 226.
20 M. Leakey, *Olduvai Gorge*, vol. 3, p. 260. Leakey also suggests the use of bolas by early hunters (see pp. 262, 266).
21 Lee, *The !Kung San*, p. 210.
22 Ibid.
23 S. L. Washburn and C. S. Lancaster, 'The Evolution of Hunting', in Lee and De Vore, *Man the Hunter*, pp. 293-303. See W. S. Laughlin, 'Hunting: an Integrating Biobehavioral System and its Evolutionary Importance', in ibid., pp. 304-20. See also C. S. Coon, *The Hunting Peoples*.

24 See E. B. Leacock, introduction to F. Engels, *The Origin of the Family, Private Property and the State.* See also E. Reed, *Woman's Evolution*, and K. Sacks, *Sisters and Wives.*
25 S. Slocum, 'Woman the Gatherer', in R. R. Reiter (ed.), *Toward an Anthropology of Women*, pp. 36-50.
26 Ibid., p. 43.
27 Ibid., p. 45.
28 Ibid., pp. 46-7.
29 Lee, *The !Kung San*, p. 262. This revises an earlier figure given by Lee which argued that gathering was 2.5 times as productive as hunting. See R. B. Lee, 'What Hunters do for a Living, or How to Make Out on Scarce Resources', in Lee and De Vore, *Man the Hunter*, p. 40.
30 See Lee, *The !Kung San*, 'Appendix B. Plants of the Dobe area', pp. 464-73.
31 Ibid., 'Appendix E. The Hand-to-Mouth Existence: a Note on the Origin of the Human Economy', pp. 489-94. See also Washburn and Lancaster, in Lee and De Vore, *Man the Hunter*, p. 297.
32 See C. E. Read-Martin and D. W. Read, 'Australopithecine Scavenging and Human Evolution: an Approach from Faunal Analysis', *Current Anthropology*, vol. 16, no. 3, 1975, pp. 359-68, and F. S. Szalay, 'Hunting–Scavenging Protohominids: a Model for Hominid Origins', *Man*, (ns), vol. 10, no. 3, 1975, pp. 420-9.
33 Isaac, 'Food-Sharing Behavior', p. 121.
34 See S. L. Washburn, discussion contribution on 'Primate Behavior and the Evolution of Aggression', in Lee and De Vore, *Man the Hunter*, p. 342.
35 M. Bicchieri, discussion contribution on 'Hunting vs. Gathering as Factors in Subsistence', ibid., p. 92.
36 Leacock, op. cit., p. 257, note 2. See G. L. Isaac, 'The Diet of Early Man: Aspects of Archaeological Evidence from Lower and Middle Pleistocene Sites in Africa', *World Archaeology*, vol. 2, no. 3, 1971, pp. 278-98, and E. J. Wing and A. B. Brown (eds), *Paleonutrition.*
37 R. A. Dart, 'The Predatory Transition from Ape to Man', *International Anthropological and Linguistic Review*, vol. 1, no. 4, 1953, p. 207.
38 R. Ardrey, *African Genesis* and *The Hunting Hypothesis.* For a review of Dart's work, see D. Wolberg, 'The Hypothesised *Osteodontokeratic* Culture of the Australopithecinae: a Look at the Evidence and the Opinions', *Current Anthropology*, vol. 11, no. 1, 1970, pp. 23-30.
39 R. A. Dart, 'The Osteodontokeratic Culture of *Australopithecus Prometheus*', *Transvaal Museum Memoir*, vol. 10, 1957.
40 Dart, 'The Predatory Transition from Ape to Man', p. 204. See also R. A. Dart, 'The Bone Tool-Manufacturing Ability of *Australopithecus Prometheus*', *American Anthropologist*, vol. 62, no. 1, 1960, pp. 134-43.
41 See Dart, 'The Predatory Transition from Ape to Man', pp. 203-4.
42 C. K. Brain, 'Some Principles in the Interpretation of Bone Accumulations Associated with Man', in G. L. Isaac and E. R. McCown (eds), *Human Origins*, p. 115.
43 C. K. Brain, comment on Wolberg, *Current Anthropology*, vol. 11, no. 1, 1970, p. 31.

44 R. Feustel, comment on Wolberg, ibid., p. 32.
45 C. K. Brain, 'New Finds at the Swartkrans Australopithecine Site', *Nature,* vol. 225, 1970, pp. 1112-19.
46 Ibid., p. 1116.
47 Ardrey, *African Genesis,* p. 300.
48 Brain, 'New Finds at the Swartkrans Australopithecine Site', p. 1118.
49 Ibid.
50 See Lee, 'What Hunters Do for a Living, or How to Make Out on Scarce Resources', in Lee and De Vore, *Man the Hunter,* pp. 30-48. Lee, *The !Kung San,* pp. 392-7, records that a particularly disruptive homicidal individual may be 'collectively executed' by members of the group and that one reason for the flexibility of the !Kung groupings is to allow conflict to be diffused by physical separation. See also J. Woodburn, 'Stability and Flexibility in Hadza Residential Groups', in Lee and De Vore, *Man the Hunter,* pp. 103-10, and C. Turnbull, contribution to discussion on 'Primate Behavior and the Evolution of Aggression', in ibid., p. 341. For a recent series of studies on this question, M. F. A. Montagu (ed.), *Learning Non-Aggression – the Experience of Pre-Literate Societies.* For an opposing view to the above, I. Eibl-Eibesfeldt, 'The Myth of the Aggression-Free Hunter and Gatherer Society', in R. Holloway (ed.), *Primate Aggression and Xenophobia,* pp. 435-57.
51 Yu. I. Semenov, 'The Doctrine of Morgan, Marxism and Contemporary Ethnography', *Soviet Anthropology and Archaeology,* vol. 4, no. 2, fall, 1965, pp. 3-15.

4 Primate tool-use and tool-making

1 For a useful review of the question of tool-use, meat-eating and 'male dominance' among primates see E. Reed, 'Primates and Prejudice', in *Sexism & Science,* pp. 8-33. See also R. S. O. Harding, 'Meat-Eating and Hunting in Baboons', in R. H. Tuttle (ed.), *Socioecology and Psychology of Primates,* pp. 245-57, and I. De Vore and S. L. Washburn, 'Baboon Ecology and Human Evolution', in F. C. Howell and F. Bourlière (eds), *African Ecology and Human Evolution,* pp. 335-67.
2 Harding, 'Meat-Eating', p. 247.
3 Ibid., p. 249.
4 Ibid., p. 250.
5 De Vore and Washburn, in Howell and Bourlière, *African Ecology,* pp. 364-5.
6 Ibid., p. 363.
7 Ibid., p. 365.
8 A. Suzuki, 'The Origin of Hominid Hunting: a Primatological Perspective', in Tuttle, *Socioecology,* p. 267.
9 Ibid., pp. 267-8.
10 Ibid., p. 261.
11 G. Teleki, 'The Omnivorous Chimpanzee', *Scientific American,* vol. 288,

no. 1, 1973, pp. 32-47. See also J. Van Lawick Goodall, *In the Shadow of Man.*

12 Teleki, 'Omnivorous Chimpanzee', p. 35, compares this form of pursuit to the kind of 'opportunistic' feeding on immobile prey that was noted for baboons.

13 Ibid., p. 37. See also R. Peters and D. L. Mech, 'Behavioral and Intellectual Adaptations of Selected Mammalian Predators to the Problem of Hunting Large Animals', in Tuttle, *Socioecology,* p. 283.

14 Teleki, op. cit., p. 35. See also J. Van Lawick Goodall, 'Continuities between Chimpanzee and Human Behavior', in G. L. Isaac and E. R. McCown (eds), *Human Origins,* p. 85.

15 Goodall, in Isaac and McCown, *Human Origins,* p. 92.

16 Teleki, op. cit., p. 37. Reed, op. cit., p. 23, argues that co-operation among primates is mainly restricted to females and their young and arises out of the female reproductive functions. Certain predator animals are capable of limited forms of co-operation when hunting. See Peters and Mech, 'Behavioral and Intellectual Adaptations', for discussion of the behaviour of wolves.

17 A. Kortlandt, discussion contribution on 'Meat-Eating and Behavioral Adaptations to Hunting' in Tuttle, *Socioecology,* p. 302. See also G. Teleki, *The Predatory Behavior of Wild Chimpanzees.*

18 Peters and Mech, op. cit., p. 288.

19 Teleki, 'The Omnivorous Chimpanzee', p. 38.

20 Ibid.

21 Ibid. See also G. L. Isaac, 'The Food-Sharing Behavior of Protohuman Hominids', *Scientific American,* vol. 238, no. 4, 1978, p. 92.

22 Teleki, 'The Omnivorous Chimpanzee', p. 42.

23 Isaac, 'Food-Sharing Behavior', p. 92.

24 Ibid., p. 113.

25 Goodall, in Isaac and McCown, *Human Origins,* p. 85.

26 Teleki, 'The Omnivorous Chimpanzee', p. 37.

27 Harding, 'Meat-Eating' p. 249.

28 Goodall, in Isaac and McCown, *Human Origins,* pp. 83-4. See also G. Teleki, 'Chimpanzee Subsistence Technology: Materials and Skills', *Journal of Human Evolution,* vol. 3, 1974, pp. 575-94.

29 Goodall, 'Tool-Using and Aimed Throwing in a Community of Free-Living Chimpanzees', *Nature,* vol. 208, 1964, pp. 1265-6.

30 J. B. Lancaster, *Primate Behavior and the Emergence of Human Culture,* p. 52.

31 Goodall, in Isaac and McCown, op. cit., p. 84.

32 G. M. Guilmet, 'The Evolution of Tool-Using and Tool-Making Behaviour', *Man* (ns), vol. 12, no. 1, 1977, pp. 38-47.

33 Ibid., p. 40.

34 G. F. Khrustov, 'The Problem of the Origin of Man', *Soviet Psychology,* vol. 9, fall, 1970, pp. 6-31.

35 Ibid., p. 28.

36 Ibid.

37 Ibid., pp. 28-29. See also P. V. Tobias, '*Australopithecus, Homo habilis,* Tool-Using and Tool-Making', *South African Archaeological Bulletin,*

vol. 20, December, 1965, pp. 167-192. Khrustov maintains, in agreement with Tobias, that australopithecines had not yet reached a level of instrumental activity beyond that of using their own natural organs and the walls and floors of caves to help shape tools. This view is not accepted by the current writer.

38 Guilmet, 'Evolution of Tool-Using', p. 41.

39 Ibid., p. 42.

40 K. P. Oakley, 'On Man's Use of Fire, with Comments on Tool-Making and Hunting', in S. L. Washburn (ed.), *Social Life of Early Man,* p. 187. See also K. P. Oakley, 'Tools Makyth Man', *Antiquity,* vol. 31, no. 124, 1957, pp. 199-209.

41 A Gruber, 'A Functional Definition of Primate Tool-Making', *Man* (ns), vol. 4, no. 4, 1969, p. 579 (original emphasis).

42 R. V. S. Wright, 'Imitative Learning of a Flake Stone Technology - the Case of an Orangutang', *Mankind,* vol. 8, no. 4, 1972, pp. 296-304.

43 Ibid., p. 298.

44 Ibid., p. 305.

45 Ibid., p. 298.

46 K. R. L. Hall, 'Tool-Using Performances as Indicators of Behavioral Adaptability', in P. C. Jay (ed.), *Primates: Studies in Adaptation and Variability,* p. 144.

47 Ibid., p. 141.

48 Ibid., p. 144.

49 B. B. Beck, *Animal Tool Behavior: The Use and Manufacture of Tools by Animals,* p. 210. Beck extends his argument on cognitive sophistication to include the utilisation of tools by early humans, thereby denying the pre-eminent role of tool-activity in shaping human development. See pp. 215-47. For a criticism of this position see T. Nishida, review, *Nature,* vol. 289, 1981, p. 616.

50 Beck, *Animal Tool Behavior,* pp. 199-214.

51 Ibid., pp. 235-236.

52 The notion of culture has also been extended to non-tool behaviour among primates. See S. Kawamura, 'The Process of Sub-Culture Propagation among Japanese Macaques', in C. H. Southwick (ed.), *Primate Social Behavior,* pp. 82-90.

53 Goodall, 'Tool-Using and Aimed Throwing in a Community of Free-Living Chimpanzees', p. 1266.

54 W. C. McGrew and C. E. G. Tutin, 'Evidence for a Social Custom in Wild Chimpanzees?', *Man* (ns), vol. 13, no. 2, 1978, pp. 234-51; and W. C. McGrew, C. E. G. Tutin and P. J. Baldwin, 'Chimpanzees, Tools and Termites: Cross Cultural Comparisons of Senegal, Tanzania and Rio Muni', *Man* (ns), vol. 14, no. 2, 1979, pp. 185-214.

55 McGrew, Tutin and Baldwin, 'Chimpanzees, Tools and Termites', pp. 205ff.

56 Ibid., p. 199 and p. 188.

57 Guilmet, 'Evolution of Tool-Using', p. 44.

58 A. Mann, 'Hominid and Culture Origins', *Man* (ns), vol. 7, no. 3, 1972, p. 382.

59 L. A. White, 'On the Use of Tools by Primates', *Journal of Comparative Psychology*, vol. 34, no. 3, 1942, pp. 369-74.
60 Ibid., p. 371.
61 McGrew and Tutin, 'Evidence for a Social Custom'.
62 White, 'Use of Tools by Primates', p. 372.
63 See S. L. Washburn and B. Benedict, 'Nonhuman Primate Culture', *Man* (ns), vol. 14, no. 1, 1979, p. 164. Goodall admits the absence of a spoken language as the main reason for the lack of progressive cultural advance among chimpanzees. See J. Van Lawick Goodall, 'Cultural Elements in a Chimpanzee Community', in E. W. Menzel (ed.) *Precultural Primate Behavior, IVth International Congress of Primatology*, vol. 1, p. 181.

5 Primate communication and culture

1 G. Mounin, 'Language, Communication, Chimpanzees', *Current Anthropology*, vol. 17, no. 1, 1976, pp. 1-21.
2 K. J. Hayes and C. H. Nissen, 'Higher Mental Functions of a Home Raised Chimpanzee', in A. M. Schrier and F. Stollnitz (eds), *Behavior of Nonhuman Primates: Modern Research Trends*, vol. 4, pp. 106-7.
3 R. A. Gardner and B. T. Gardner, 'Teaching Sign Language to a Chimpanzee', *Science*, vol. 165, 1969, pp. 664-72.
4 B. T. Gardner and R. A. Gardner, 'Two-Way Communication with an Infant Chimpanzee', in Schrier and Stollnitz, *Behavior of Nonhuman Primates*, pp. 147-57.
5 Ibid., pp. 168-78.
6 D. Premack, 'Language in Chimpanzee?' *Science*, vol. 172, 1971, pp. 808-22. See D. Premack, *Intelligence in Ape and Man*, for a full report of these researches.
7 D. Premack, 'On the Assessment of Language Competence in the Chimpanzee', in Schrier and Stollnitz, *Behavior of Nonhuman Primates*, pp. 185-228.
8 D. M. Rumbaugh, E. C. Von Glasersfeld, T. V. Gill, H. Warner, P. Pisani, J. V. Brown and C. L. Bell, 'The Language Skills of a Young Chimpanzee in a Computer-Controlled Training Situation', in R. H. Tuttle (ed.), *Socioecology and Psychology of Primates*, pp. 391-401. See D. M. Rumbaugh (ed.), *Language Learning by a Chimpanzee: The Lana Project*, for a full account of this work.
9 R. S. Fouts, 'Capacities for Language in Great Apes', in Tuttle, *Socioecology*, p. 380.
10 H. S. Terrace, *Nim*. This work was originally undertaken to challenge Chomsky's view of the 'innateness' of human language, hence Neam Chimpsky, (Nim). See also H. S. Terrace, L. A. Petitto, R. J. Saunders and T. G. Bever 'Can an Ape Create a Sentence?', *Science*, vol. 206, 1979, pp. 891-906.
11 Ibid., p. 892.
12 Ibid., p. 893.

13 Ibid., p. 894.
14 Ibid.
15 Ibid., p. 895.
16 Fouts, in Tuttle, *Socioecology*, p. 382.
17 Terrace, Pettito, Saunders and Bever, 'Can an Ape Create a Sentence', pp. 896-7.
18 Ibid., p. 896.
19 Ibid., p. 897.
20 Ibid.
21 Ibid., pp. 898-9.
22 Ibid., p. 900.
23 Ibid., pp. 900-1.
24 Premack's study of Sarah has also been subject to a careful critique along similar lines by Terrace. See H. S. Terrace, 'Is Problem-Solving Language?', in T. A. Sebeok and J. Umiker-Sebeok (eds), *Speaking of Apes: A Critical Anthology of Two-Way Communication with Man*, pp. 385-405.
25 See C. R. Thompson and R. M. Church, 'An Explanation of the Language of a Chimpanzee', *Science*, vol. 208, 1980, pp. 313-14.
26 Ibid., p. 314.
27 Extensive reviews of this field of research may be found in J. R. Oppenheimer, 'Communication in New World Monkeys'; J. P. Gautier and A. Gautier, 'Communication in Old World Monkeys'; and P. Marler and R. Tenaza, 'Signalling Behavior of Apes with Special Reference to Vocalization', in T. A. Sebeok (ed.), *How Animals Communicate*, pp. 851-89, pp. 890-964 and pp. 965-1033 respectively.
28 J. B. Lancaster, 'Primate Communication Systems and the Emergence of Human Language', in P. C. Jay (ed.), *Primates: Studies in Adaptation and Variability*, p. 442.
29 Ibid., p. 442.
30 See S. A. Altmann, 'The Structure of Primate Social Communication', in S. A. Altmann (ed.), *Social Communication among Primates*, pp. 325-62.
31 P. Marler, 'Communication in Monkeys and Apes', in I. De Vore (ed.), *Primate Behavior*, pp. 544-84.
32 See T. T. Struhsaker, 'Auditory Communication among Vervet Monkeys *(Cercopithecus aethiops)*', in Altmann, *Social Communication*, pp. 281-324. Struhsaker discusses thirty-six discrete sounds among vervet monkeys. Struhsaker has claimed that differentiated vocalisations are linked with either a mammal predator, the presence of a snake or a human being, i.e., that they convey specific information about the kind of danger in the environment. Gautier and Gautier, 'Communication in Old World Monkeys', p. 909, suggest, however, that instead of specific reactions, differences in the levels of excitation produce responses of varied intensity. For comments on the social and environmental context of primate signals see Gautier and Gautier, op. cit., pp. 955-8.
33 J. B. Lancaster, *Primate Behavior and the Emergence of Human Culture*, p. 58.
34 Marler, op. cit., p. 584.

35 For an opposing view see E. S. Savage-Rumbaugh, D. M. Rumbaugh and S. Boyer, 'Linguistically Mediated Tool Use and Exchange by Chimpanzees (Pan Troglodytes)', *The Behavioral and Brain Science,* vol. 4, 1978, pp. 539-54 reprinted in Sebeok and Umiker-Sebeok, *Speaking of Apes,* pp. 353-83.
36 Thus Loy claims on the basis of his study of Japanese macaques that hierarchy systems in human culture are an expression of the 'primate biogram' and to the extent that australopithecines exhibited dominance they had begun the transition toward 'the cultural status arrangements of true man'. See J. Loy, 'The Descent of Dominance in *Macaca:* Insights into the Structure of Human Societies', in R. H. Tuttle (ed.), *Socioecology and Psychology of Primates,* pp. 153-80.
37 E. O. Wilson, *Sociobiology: The New Synthesis.*
38 See in addition to works cited below, Sociobiology Study Group of Science for the People, 'Sociobiology: Another Biological Determinism', *BioScience,* vol. 26, no. 3, 1976, pp. 182-6, reprinted in A. L. Caplan (ed.), *The Sociobiology Debate,* pp. 280-90. See also M. F. A. Montagu (ed.), *Sociobiology Examined.*
39 E. O. Wilson, *On Human Nature.*
40 For a description of the genetic theory underlying sociobiology see M. Ruse, *Sociobiology: Sense or Nonsense?*
41 Wilson, *On Human Nature,* p. 85.
42 This argument is pursued by Barkow in an attempt to defend sociobiology. See J. H. Barkow, 'Culture and Sociobiology', *American Anthropologist,* vol. 80, no. 1, 1980, pp. 5-20.
43 T. Dobzhansky, 'Cultural Direction of Human Evolution', *Human Biology,* vol. 35, no. 3, 1963, pp. 311-16.
44 Wilson, *On Human Nature,* p. 86.
45 Ibid., p. 89.
46 Ibid., p. 97.
47 See M. Sahlins, *The Use and Abuse of Biology.* See also the critique by Science as Ideology Group of the British Society for Social Responsibility in Science, 'The New Synthesis is an Old Story', *New Scientist,* vol. 70, no. 1000, 1976, pp. 346-8.
48 S. J. Gould, 'Sociobiology: the Art of Storytelling', *New Scientist,* vol. 80, 1978, pp. 530-3.
49 Ibid., p. 532.
50 Ibid.
51 Ibid.
52 S. L. Washburn, 'Human Behavior and the Behavior of Other Animals', *American Psychologist,* vol. 33, May, 1978, pp. 405-18.
53 Ibid., p. 411. Social learning is important in the behaviour of animals other than humans but, as previously suggested, such learning tends to be non-cumulative. Modification of learned behavior and its transmission to subsequent generations is relatively infrequent. In general the existing stock of learned behaviour is simply perpetuated.
54 T. Dobzhansky, *Mankind Evolving,* p. 319.

55 T. Dobzhansky, *The Biology of Ultimate Concern*, p. 43.
56 Sahlins, *Use and Abuse*, pp. 61-5.
57 Ibid., p. 13.

6 Theories of language origins

1 For a review of the historical theories of language origins see G. W. Hewes, 'Language Origin Theories' in D. M. Rumbaugh (ed.), *Language Learning by a Chimpanzee: The Lana Project*, pp. 3-53.
2 C. Hockett, 'Animal "Languages" and Human Language', in J. N. Spuhler (ed.), The Evolution of Man's Capacity for Culture, *Human Biology*, vol. 31, no. 1, 1959, p. 32.
3 C. Hockett, 'The Origin of Speech', *Scientific American*, vol. 203, no. 3, 1960, pp. 88-96.
4 Ibid., pp. 90-1.
5 Ibid., pp. 94-5.
6 Hockett in Spuhler, op. cit., p. 36. Hockett, however, does not view displacement as peculiar to human communication, arguing that the dance of the bees, signalling the location of food sources is an instance of the same phenomenon.
7 Hockett lists 'cultural transmission' as a separate design feature. It is not clear therefore whether each of the properties listed has equal explanatory weight.
8 G. W. Hewes, 'Primate Communication and the Gestural Origin of Language', *Current Anthropology*, vol. 14, nos 1-2, 1973, pp. 5-12.
9 Ibid., p. 6.
10 R. Fouts, 'Language: Origin, Definition and Chimpanzees', *Journal of Human Evolution*, vol. 3, 1974, pp. 475-82.
11 Hewes, 'Primate Communication', p. 7.
12 Ibid., p. 8.
13 Ibid.
14 Ibid., p. 9.
15 Ibid., p. 10.
16 P. Lieberman, E. S. Crelin and B. M. Klatt, 'Phonetic Ability and Related Anatomy of the Newborn and Adult Human, Neanderthal Man, and the Chimpanzee', *American Anthropologist*, vol. 74, no. 3, 1972, pp. 287-307. For an opposing view see J. Wind, 'Phylogeny of the Human Vocal Tract', in S. R. Harnad, H. D. Steklis and J. Lancaster (eds), *Origins and Evolution of Language and Speech*, New York Academy of Sciences, vol. 280, 1976, pp. 612-30.
17 There is an extensive literature on non-verbal communication. See S. Weitz (ed.), *Nonverbal Communication*, and A. Wolfgang (ed.), *Nonverbal Behavior: Applications and Cultural Implications*.
18 Hewes, 'Primate Communication and the Gestural Origin of Language', p. 10.
19 Ibid., pp. 9-10.
20 C. Darwin, *The Descent of Man*, p. 55.

21 A. I. Hallowell, 'The Protocultural Foundations of Human Adaptation', in S. L. Washburn (ed.), *Social Life of Early Man*, p. 248.
22 A. I. Hallowell, 'Self, Society and Culture in Phylogenetic Perspective', in S. Tax (ed.), *Evolution after Darwin*, vol. 2, p. 357.
23 Ibid., p. 346.
24 Ibid., pp. 348-59.
25 Ibid., p. 347. For a non-psychoanalytic approach to the question of incest taboo, see L. White, *The Evolution of Culture*.
26 Hallowell, 'Self, Society and Culture in Phylogenetic Perspective', p. 319.
27 Ibid., p. 344.
28 Ibid., p. 357.
29 R. L. Holloway, 'Culture: a *Human* Domain', *Current Anthropology*, vol. 10, no. 4, 1969, pp. 395-412.
30 Ibid., pp. 396-7.
31 Ibid., p. 395.
32 Ibid., p. 401.
33 Ibid., p. 399.
34 Ibid.
35 Ibid., p. 401. It has already been argued that 'standardisation' may be a poor guide to the complexity of the work processes underlying tool-making. See chapter 2, p. 17.
36 Ibid.
37 Ibid., pp. 402-4.
38 Ibid., p. 406.
39 Ibid., pp. 403-4.
40 Ibid., p. 404.
41 See R. N. Bowen, comment on Holloway, ibid., p. 408.
42 J. B. Lancaster, 'Primate Communication Systems and the Emergence of Human Language', in P. C. Jay (ed.), *Primates: Studies in Adaptation and Variability*, pp. 439-57.
43 Ibid., p. 440.
44 Ibid., p. 454.
45 Ibid., p. 456.
46 Ibid., p. 452. Lancaster emphasises the importance of the angular gyrus. Certain capacity for cross-modal association has however been shown for great apes. See R. K. Davenport, 'Cross-Modal Perception in Apes', in Harnad, Steklis and Lancaster, *Origins and Evolution*, pp. 143-9. Interestingly, the poorest performance was recorded for those cross-modal transfers requiring generalisation of haptic (touch) to visual perception.
47 Lancaster, 'Primate Communication', pp. 447-54. See also B. W. Robinson, 'Vocalization Evoked from Forebrain in *Macaca mulatta*', *Physiology and Behavior*, vol. 2, no. 4, 1967, pp. 345-54.
48 Ibid., p. 454.
49 Ibid., pp. 454-9.
50 J. Limber, 'Language in Child and Chimp?', in T. A. Sebeok and J. Umiker-Sebeok (eds), *Speaking of Apes: A Critical Anthology of*

Two-way Communication with Man, pp. 197-220.
51 Ibid., p. 200.
52 L. S. Vygotsky, *Thought and Language.*
53 Ibid., p. 6.
54 L. Noiré, *The Origin and Philosophy of Language*, p. 143. There is no evidence that either Noiré or Engels knew of the other's work.
55 Ibid., p. 139.
56 Ibid., p. 138. Noiré also stressed the act of naming and 'action' words and can be held to prefigure both Lancaster's and Hewes's views on this. Noiré has also been credited with influencing the theories of N. Y. Marr and with the invention of the 'work-chant' theory. See G. Thomson, *The Human Essence*, for a more recent suggestion that the syntactical structure of language directly reflects the subject/object relations of labour process.
57 G. A. De Laguna, *Speech: Its Function and Development*, p. 19.
58 Ibid.
59 Ibid.
60 Ibid.
61 G. Révész, *The Origins and Prehistory of Language*, p. 114. Like Noiré, action words or verbs connected with labour are seen as the basic elements of language. The notion of 'contact', however, rests on an innatist conception.
62 Ibid., p. 93.
63 Ibid.
64 A. Montagu, 'Toolmaking, Hunting and the Origin of Language', in Harnad, Steklis and Lancaster, *Origins and Evolution*, pp. 266-74.
65 Ibid., p. 267.
66 Although food-gathering is mentioned it is not perhaps sufficiently stressed.
67 Ibid., p. 269.
68 Ibid. Holloway's notion of a 'grammar' of tool-making is also cited in this regard.
69 Ibid.
70 V. V. Bunak, 'Present State of the Problem of the Origin of Speech and the Early Stages of its Evolution', *Journal of World History*, vol. 5, no. 2, 1959, pp. 310-24.

7 Labour and culture

1 The following discussion of Soviet psychology is taken in part from an article by the current writer entitled 'Culture, Language and the Human Personality', *Marxism Today*, vol. 21, no. 8, 1977, pp. 229-40. Further expositions of Pavlov's work on speech, in addition to those cited below, are contained in P. K. Anokhin, 'Ivan P. Pavlov and Psychology', in B. Wolman (ed.), *Historical Roots of Contemporary Psychology*, pp. 131-59, and F. E. X. Dance, 'Speech Communication Theory and Pavlov's Second Signal System', *Journal of Communication*,

vol. 17, 1967, pp. 13-24. For a survey of Pavlov's more general work see J. A. Gray, *Pavlov.*

2 I. P. Pavlov, *Selected Works,* pp. 651-2.

3 Cited in D. G. Bogoiavlensky, 'The Psychology of Understanding', in B. Simon (ed.), *Psychology in the Soviet Union,* p. 75.

4 Pavlov's conception of verbal activity and conditioned reflex behaviour should be sharply distinguished from mechanistic behaviourist approaches. See O. K. Tikhomirov, review of B. F. Skinner, *Verbal Behavior, Word,* vol. 15, 1959, pp. 362-7, and K. F. Riegel, 'Structure and Transformation in Modern Intellectual History', in K. F. Riegel and G. C. Rosenwald (eds), *Structure and Transformation: Developmental and Historical Aspects,* pp. 3-45.

5 Tikhomirov, Review of Skinner, p. 366.

6 Bogoiavlensky, 'Psychology of Understanding', p. 74.

7 A remarkable plea for Western social scientists to incorporate the valuable legacy of historical materialist psychology left by Vygotsky and Luria has recently been made by Stephen Toulmin of Chicago. See S. Toulmin, 'The Mozart of Psychology', *New York Review of Books,* vol. 25, no. 14, September, 1978, pp. 51-7.

8 See A. R. Luria, *The Role of Speech in the Regulation of Normal and Abnormal Behaviour.*

9 Ibid., p. 22. For further discussion of the notion of inner speech as elaborated by Vygotsky, see A. A. Leontiev, 'Some New Trends in Soviet Psycholinguistics', *Soviet Psychology,* vol. 25, no. 2, winter, 1976, pp. 15-25.

10 A. R. Luria, 'The Role of Language in the Formation of Temporary Connections', in Simon, *Psychology in the Soviet Union,* p. 112.

11 Luria, *The Role of Speech in the Regulation of Normal and Abnormal Behaviour,* p. 23.

12 L. Rahmani, *Soviet Psychology,* p. 42. Luria has thus been highly critical of the 'innatist' view of language acquisition put forward by Chomsky. See A. R. Luria, 'Scientific Perspectives and Philosophical Dead Ends in Modern Linguistics', *Cognition,* vol. 3, no. 4, 1974-5, pp. 377-85. For Chomsky's reply, see N. Chomsky, 'On the Biological Basis of Language Capacities', in R. W. Rieber (ed.), *The Neuropsychology of Language: Essays in Honor of Eric Lenneberg,* pp. 1-24.

13 Apart from Vygotsky's own work, useful commentaries are contained in A. N. Leontiev and A. R. Luria, 'The Psychological Ideas of L. S. Vygotskii', in Wolman, *Historical Roots,* pp. 338-67; L. I. Bozhovich, 'The Concept of the Cultural-Historical Development of the Mind and its Prospects', *The Soviet Review,* vol. 19, no. 1, spring, 1980, pp. 31-48; and the important collection of articles in the Vygotsky Memorial Issue, *Soviet Psychology,* vol. 5, no. 3, fall, 1967, introduced by J. S. Bruner. An early attempt to apply some of Vygotsky's insights by a non-Marxist palaeoanthropologist is A. Roe, 'Psychological Definitions of Man', in S. L. Washburn (ed.), *Classification and Human Evolution,* pp. 320-31.

14 L. S. Vygotsky, 'The Development of Higher Psychological Functions',

The Soviet Review, vol. 18, no. 3, fall, 1977, p. 43 (original emphasis). A new edition of Vygotsky's writings in English edited by Cole *et al.* contains this essay under the title 'Internalization of Higher Psychological Functions'. See L. S. Vygotsky, *Mind in Society: The Development of Higher Psychological Processes*, pp. 52-7. It should be noted, however, that the first pages of the original translation are excluded from the new edition, including the quotation from Pavlov's work, while two further pages are added at the end of the article. *The Soviet Review* version of the article is not referred to in Cole's bibliography of Vygotsky's work in English, although he edited this version as well.

15 A. A. Leontiev, 'Sense as a Psychological Concept', in J. Prucha (ed.), *Soviet Studies in Language and Language Behaviour*, pp. 82-90.

16 Vygotsky, 'The Development of Higher Psychological Functions', p. 49.

17 Ibid., p. 50 (original emphasis).

18 Ibid., p. 51.

19 See L. S. Vygotsky, 'Tool and Symbol in Child Development', in Vygotsky, *Mind in Society: The Development of Higher Psychological Processes*, pp. 19-30. It is interesting to note that Vygotsky observes that speech follows action and only at a later stage of development moves to the starting-point of activity. Ibid., pp. 27-8. Vygotsky's work also shows the futility of those arguments which attempt artificially to separate 'technical thinking' or 'thinking in terms of tools' from language.

20 See also L. S. Vygotsky, *Thought and Language*.

21 See F. Engels, *Dialectics of Nature*, p. 172:

> Both natural science and philosophy have until now, absolutely ignored the study of the effect of man's activity on his mental processes. They consider, on the one hand, only nature and, on the other, only thought. The most *essential* basis of human thought consists in the *changing of nature by man*, not simply in nature as such; human reason has developed in accordance with man's alteration of nature.

Cited in Vygotsky, 'The Development of Higher Psychological Functions', p. 43 (original emphasis). At the time Lenin turned his attention to the question of dialectical materialism in 1908 in his *Materialism and Empirio-Criticism*, in which he presents a vigorous defence of Engels's writings on dialectics, he was unfortunately not aware of the existence of *Dialectics of Nature*.

22 J. McLeish, *Soviet Psychology: History, Theory, Content*, p. 144.

23 The implications of the Pavlovian second-signalling system for the theory of reflection and its relevance for Engels's theory of human development is discussed in H. K. Wells, *Ivan P. Pavlov: Toward a Scientific Psychology and Psychiatry*, pp. 72-87.

24 See A. Spirkin, 'Consciousness', in T. J. Blakeley (ed.), *Themes in Soviet Marxist Philosophy*, pp. 59-73.

25 V. I. Lenin, *Philosophical Notebooks*, Collected Works, vol. 38, p. 274 (original emphasis).

26 B. F. Lomov, 'Lenin's Theory of Reflection and Psychology', *Studia*

Psychologica, vol. 13, no. 3, 1971, p. 176. See also D. Goldstick, 'The Leninist Theory of Perception', *Dialogue*, vol. 19, no. 1, 1980, pp. 1-19.

27 A. N. Leontiev and A. A. Leontiev, 'The Social and the Individual in Language', *Language and Speech*, vol. 2, no. 4, 1959, p. 191.

28 The preliminary working out of the complex character of objective activity structures is probably the most important advance of contemporary Marxist psychology. See especially A. N. Leontiev, 'The Problem of Activity in Psychology', *Soviet Psychology*, vol. 13, no. 2, winter, 1974-5, pp. 4-33, and A. N. Leontiev, *Activity, Consciousness and Personality*.

39 K. Marx, *The German Ideology*, p. 51.

30 Ibid.

31 G. Lukacs, *History and Class Consciousness*, preface to new edition, 1967, p. xxv. Cf. J. Hoffman, *Marxism and the Theory of Praxis*, pp. 70-108.

32 V. Gordon Childe, 'Prehistory and Marxism', *Antiquity*, vol. 53, no. 208, 1979, pp. 93-5.

33 K. Marx, *Grundrisse*, p. 123.

34 E. V. Ilyenkov, 'The Concept of the Ideal', in *Philosophy in the USSR: Problems of Dialectical Materialism*, pp. 71-99.

35 T. Dobzhansky, *The Biology of Ultimate Concern*, p. 56.

36 M. Critchley, 'The Evolution of Man's Capacity for Language', in S. Tax (ed.), *Evolution after Darwin*, vol. 2, p. 378.

37 Ibid., p. 297. See R. Julkunen, 'A Contribution to the Categories of Social Time and the Economy of Time', *Acta Sociologica*, vol. 20, no. 1, 1977, pp. 5-24. The notion of 'economy of time' should be distinguished from 'rhythm of life' as the subjective form of the early consciousness of time directly 'imposed' by nature. The allocation of time need not be conscious except in class society where there is a contradiction between free time and labour time.

38 Marx, *Grundrisse*, p. 711.

39 L. White, *The Science of Culture*, p. 369.

40 M. Sahlins, *Stone Age Economics*, pp. 1-39.

41 Engels, *Dialectics of Nature*, p. 261.

42 K. Marx, letter to Ludwig Kugelmann in Hanover, 11 July 1868, in K. Marx and F. Engels, *Selected Correspondence*, pp. 195-7.

43 Anokhin, 'Pavlov and Psychology'.

44 See V. I. Kochetkova, *Paleoneurology*, especially pp. 247-306.

45 Ibid., p. 260. On fossil endocasts see also chapter 2, note 43. The 'frontal and parieto-temporal field complexes' are noted as the phylogenetically new areas of the brain. Kochetkova's work is based on that of Luria. See A. R. Luria, *The Working Brain*. For criticisms of Luria's model see B. Kolb and I. Q. Whishaw, *Fundamentals of Human Neuropsychology*, pp. 145-6. See also the important collection of articles published in *The Brain*, *Scientific American*, vol. 241, no. 9, 1979.

Kochetkova locates the emergence of culture in the later periods along with more complex tool-making which is given detailed analysis. The view of Oakley, however, suggests that the making of Oldowan tools, involving flaking in more than one direction to produce a point or a cutting edge, was already becoming a highly skilful process requiring 'much foresight', complex neuro-muscular patterns and some form of communication. This suggests a much earlier embedding of human labour activity in culture. See K. P. Oakley, 'Man the Skilled Tool-maker', *Antiquity,* vol. 43, no. 171, 1969, pp. 222-4.

46 B. Campbell, 'The Roots of Language', in J. Morton (ed.), *Biological and Social Factors in Psycholinguistics,* p. 17.

47 Kochetkova's reliance on endocast evidence to some extent leads her to view only larger-brained hominids as 'human'.

48 Engels, *Dialectics of Nature,* p. 263. A recent review of the philosophical underpinnings of evolutionary theory, which attempts to situate Darwin and post-Darwinian palaeontology in terms of the long-running debate between gradualistic and discontinuous approaches to evolutionary change, is contained in J. Cracraft, 'Phylogenetic Analysis, Evolutionary Models, and Paleontology, pp. 7-39. A contemporary example of the latter approach is I. Tattersall and N. Eldredge, 'Fact, Theory and Fantasy in Human Paleontology', *American Scientist,* vol. 65, no. 2, 1977, pp. 204-11. For an alternative view see J. E. Cronin, N. T. Boaz, C. B. Stringer and Y. Rak, 'Tempo and Mode in Hominid Evolution', *Nature,* vol. 292, 1981, pp. 113-22.

The idea of evolutionary change occurring in fairly 'rapid' sequences interspersing much longer periods of relative equilibrium is by no means entirely new. Thomas Huxley is referred to by Gould as having tackled Darwin on this very point. See R. E. F. Leakey, 'The Making of Mankind', the *Listener,* vol. 105, no. 2711, May 7 1981, p. 600. Since then, an impressive range of evolutionary biologists and palaeontologists have challenged the postulate of evolutionary gradualism, especially at the level of 'macroevolution'. See R. Lewin, 'Evolutionary Theory under Fire', *Science,* vol. 210, 1980, pp. 883-7. Dobzhansky has given a useful general formulation of discontinuity: 'development does not flow at a uniform rate; from time to time it involves apparent breaks of the continuity, giving rise to something radically new' (T. Dobzhansky, comment on R. L. Holloway, 'Culture: a *Human* Domain', *Current Anthropology,* vol. 10, no. 4, 1969, p. 409).

49 S. L. Washburn, 'Speculations on the Interrelations of the History of Tools and Biological Evolution', in J. N. Spuhler (ed.), The Evolution of Man's Capacity for Culture, *Human Biology,* vol. 31, no. 1, 1959, p. 29.

50 See Engels, *Dialectics of Nature,* p. 18:

> But step by step with the development of the hand went that of the brain; first of all consciousness of the conditions for separate practically useful actions, and later, among the more favoured peoples and arising from the preceding, insight into the natural laws governing them. And with the rapidly growing knowledge of the laws of nature the means for reacting on nature also grew; the hand alone would never have achieved the steam engine if the brain of man had not attained a correlative development with it, and parallel to it, and partly owing to it.

51 S. L. Washburn and J. B. Lancaster, 'On Evolution and the Origin of Language', *Current Anthropology,* vol. 12, no. 3, 1971, p. 384. See also W. Howells, *Evolution of the Genus 'Homo',* p. 48. The absence of such progressive neuro-sensory links in anthropoids is suggested by Davenport's work. See comment in chapter 6, note 46.
52 See W. Penfield and T. Rasmussen, *The Cerebral Cortex of Man,* pp. 214-15.
53 A. A. Leontiev, 'Social and Natural in Semiotics', in Morton, *Biological and Social Factors,* p. 124.
54 G. F. Debetz, 'The Social Life of Early Paleolithic Man as Seen Through the Work of Soviet Anthropologists', in S. L. Washburn (ed.), *Social Life of Early Man,* pp. 137-49.
55 Engels, *Dialectics of Nature,* p. 289.
56 Ibid.
57 See E. B. Leacock, editor's introduction to F. Engels, 'The Part Played by Labor in the Transition from Ape to Man', in F. Engels, *The Origin of the Family, Private Property and the State,* p. 289.

Bibliography

Altmann, S. A. (ed.), *Social Communication among Primates*, University of Chicago Press, 1967.

Altmann, S. A., 'The Structure of Primate Social Communication', in Altmann (ed.), 1967, pp. 325-62.

Andrews, P., 'Hominoid Habitats of the Miocene', *Nature*, vol. 289, 1981, p. 749.

Andrews, P., and Tobein, H., 'New Miocene Locality in Turkey with Evidence on the Origin of *Ramapithecus* and *Sivapithecus*', Nature, vol. 268, 1977, pp. 699-701.

Andreyev, I., 'The Origins of Man and Society', *Social Sciences*, USSR, vol. 8, no. 3, 1977, pp. 101-12.

Anokhin, P. K., 'Ivan P. Pavlov and Psychology', in Wolman (ed.), 1968, pp. 131-59.

Ardrey, R., *African Genesis*, London, Collins, 1961.

Ardrey, R., *The Hunting Hypothesis*, London, Collins, 1976.

Ball, T., 'Marx and Darwin: A Reconsideration', *Political Theory*, vol. 7, no. 4, 1979, pp. 469-84.

Barkow, J. H., 'Culture and Sociobiology', *American Anthropologist*, vol. 80, no. 1, 1980, pp. 5-20.

Beck, B. B., *Animal Tool Behavior: The Use and Manufacture of Tools by Animals*, New York, Garland STPM Press, 1980.

Bernal, J. D., *The Freedom of Necessity*, London, Routledge & Kegan Paul, 1949.

Bicchieri, M., discussion contribution on 'Hunting vs. Gathering as Factors in Subsistence', in Lee and De Vore (eds), 1968, p.92.

Binford, S. R., 'Ethnographic Data and Understanding the Pleistocene', in Lee and De Vore (eds), 1968, pp. 274-5.

Binford, S. R., and Binford, L. R., 'Stone Tools and Human Behavior', *Scientific American*, vol. 220, no. 4, 1969, pp. 70-84.

Blakeley, T. J. (ed.), *Themes in Soviet Marxist Philosophy*, Dordrecht, Holland, Reidel, 1975.

Bogoiavlensky, D. G., 'The Psychology of Understanding', in Simon (ed.), 1957, pp. 69-83.

Bowen, R. N., comment on Holloway, 1969, *Current Anthropology*, vol. 10, no. 4, 1969, p. 408.

Bozhovich, L. I., 'The Concept of the Cultural-Historical Development of the Mind and its Prospects', *The Soviet Review,* vol. 19, no. 1, spring, 1980, pp. 31-48.

Brace, C. L., and Montagu, A., *Human Evolution,* New York, Macmillan, 1977.

Brain, C. K., 'New Finds at the Swartkrans Australopithecine Site', *Nature,* vol. 225, 1970, pp. 1112-19.

Brain, C. K., comment on Wolberg, *Current Anthropology,* vol. 11, no. 1, 1970, pp. 31-2.

Brain, C. K., 'Some Principles in the Interpretation of Bone Accumulations Associated with Man', in Isaac and McCown (eds), 1976, pp. 97-116.

Bunak, V. V., 'Present State of the Problem of the Origin of Speech and the Early Stages of its Evolution', *Journal of World History,* vol. 5, no. 2, 1959, pp. 310-24.

Campbell, B., 'The Roots of Language', in Morton (ed.), 1971, pp. 10-23.

Caplan, A. L. (ed.), *The Sociobiology Debate,* New York, Harper & Row, 1978, pp. 280-290.

Carver, T., 'Marx, Engels and Dialectics', *Political Studies,* vol. 28, no. 3, 1980, pp. 353-63.

Childe, V. Gordon, 'Prehistory and Marxism', *Antiquity,* vol. 53, no. 208, 1979, pp. 93-5.

Chomsky, N., 'On the Biological Basis of Language Capacities', in Rieber (ed.), 1976, pp. 1-24.

Clark, J. D., 'African Origins of Man the Tool-Maker', in Isaac and McCown (eds), 1976, pp. 1-53.

Coon, C. S., *The Hunting Peoples,* Harmondsworth, Penguin, 1976.

Coppens, Y., Howell, F. C., Isaac, G. L. and Leakey, R. E. F. (eds), *Earliest Man and Environments in the Lake Rudolf Basin,* University of Chicago Press, 1976.

Coursey, D. S., 'Hominid Evolution and Hypogeous Plant Foods', *Man* (ns), vol. 8, no. 4, 1973, pp. 364-5.

Cracraft, J., 'Phylogenetic Analysis, Evolutionary Models and Paleontology', in Cracraft and Eldredge (eds), 1979, pp. 7-39.

Cracraft, J., and Eldredge, N. (eds), *Phylogenetic Analysis and Paleontology,* Columbia University Press, 1979.

Critchley, M., 'The Evolution of Man's Capacity for Language', in Tax (ed), 1960, pp. 289-301.

Cronin, J. E., Boaz, N. T., Stringer, C. B. and Rak, Y., 'Tempo and Mode in Hominid Evolution', *Nature,* vol. 292, 1981, pp. 113-22.

Crosby, J. F., 'Evolutionism and the Ontology of the Human Person: a Critique of the Marxist Theory of the Emergence of Man', *Review of Politics,* vol. 38, no. 2, 1976, pp. 208-43.

Dance, F. E. X., 'Speech Communication in the Soviet Union: the Phylogenesis of Speech According to Frederick Engels', *The Speech Teacher, Journal of Speech Association of America,* vol. 13, 1964, pp. 115-18.

Dance, F. E. X., 'Speech Communication Theory and Pavlov's Second Signal System', *Journal of Communication,* vol. 17, 1967, pp. 13-24.

Dart, R. A., 'The Predatory Transition from Ape to Man', *International Anthropological and Linguistic Review,* vol. 1, no. 4, 1953, pp. 201-19.

Dart, R. A., 'The Osteodontokeratic Culture of *Australopithecus Prometheus*', *Transvaal Museum Memoir*, vol. 10, 1957.

Dart, R. A., 'The Bone Tool-Manufacturing Ability of *Australopithecus Prometheus*', *American Anthropologist*, vol. 62, no. 1, 1960, pp. 134-43.

Darwin, C., *The Descent of Man*, London, Watts, 1930.

Davenport, R. K., 'Cross-Modal Perception in Apes', in Harnad, Steklis and Lancaster (eds), 1976, pp. 143-9.

Day, M., 'Hominid Postcranial Material from Bed I, Olduvai Gorge', in Isaac and McCown (eds), 1976, pp. 363-74.

Debetz, G. F., 'The Social Life of Early Paleolithic Man as Seen Through the Work of Soviet Anthropologists', in Washburn (ed.), 1962, pp. 137-49.

De Laguna, G. A., *Speech: Its Function and Development*, Indiana University Press (1929), 1963.

De Vore, I. (ed.), *Primate Behavior*, New York, Holt, Rinehart & Winston, 1965.

De Vore, I., and Washburn, S. L., 'Baboon Ecology and Human Evolution', in Howell and Bourlière (eds), 1964, pp. 353-67.

Diamond, S. (ed), *Towards a Marxist Anthropology: Problems and Perspectives*, The Hague, Mouton, 1979.

Dobzhansky, T., *Mankind Evolving*, Yale University Press, 1962.

Dobzhansky, T., 'Cultural Direction of Human Evolution', *Human Biology*, vol. 35, no. 3, 1963, pp. 311-16.

Dobzhansky, T., *The Biology of Ultimate Concern*, London, Rapp & Whiting, 1969.

Dobzhansky, T., comment on Holloway, 1969, *Current Anthropology*, vol. 10, no. 4, 1969, p. 409.

Eibl-Eibesfeldt, I., 'The Myth of Aggression-Free Hunter and Gatherer Society', in Holloway (ed), 1974, pp. 435-57.

Engels, F., *Dialectics of Nature*, London, Lawrence & Wishart, 1940.

Engels, F., 'The Part Played by Labour in the Transition from Ape to Man', in Engels, 1940, pp. 279-96.

Engels, F., *The Origin of the Family, Private Property and the State*, London, Lawrence & Wishart, 1972.

Feustel, R., comment on Wolberg, *Current Anthropology*, vol. 11, no. 1, 1970, p. 32.

Fouts, R. S., 'Language: Origin, Definition and Chimpanzees', *Journal of Human Evolution*, vol. 3, 1974, pp. 475-82.

Fouts, R. S., 'Capacities for Language in Great Apes', in Tuttle (ed), 1975, pp. 371-90.

Freeman, L. B., 'A Theoretical Framework for Interpreting Archaeological Materials', in Lee and De Vore (eds), 1968, pp. 264-5.

Gardner, B. T., and Gardner, R. A., 'Two-Way Communication with an Infant Chimpanzee', in Schrier and Stollnitz (eds), 1971, pp. 117-84.

Gardner, R. A., and Gardner, B. T., 'Teaching Sign Language to a Chimpanzee', *Science*, vol. 165, 1969, pp. 664-72.

Gautier, J. P., and Gautier, A., 'Communication in Old World Monkeys', in Sebeok (ed), 1977, pp. 890-946.

Geertz, C., 'The Transition to Humanity', in Tax (ed), 1965, pp. 37-48.

Goldstick, D., 'The Leninist Theory of Perception', *Dialogue*, vol. 19, no. 1, 1980, pp. 1-19.

Goodall, J., 'Tool-Using and Aimed Throwing in a Community of Free-Living Chimpanzees', *Nature*, vol. 208, 1964, pp. 1265-6.

Goodall, J., 'Cultural Elements in a Chimpanzee Community', in Menzel (ed), 1973, pp. 144-184.

Goodall, J., *In the Shadow of Man*, Glasgow, Collins, 1976.

Goodall, J., 'Continuities between Chimpanzee and Human Behavior', in Isaac and McCown (eds), 1976, pp. 81-95.

Gould, S. J., 'Sociobiology: the Art of Storytelling', *New Scientist*, vol. 80, 1978, pp. 530-3.

Gray, J. A., *Pavlov*, Glasgow, Fontana, 1979.

Gruber, A., 'A Functional Definition of Primate Tool-Making', *Man* (ns), vol. 4, no. 4, 1969, pp. 537-79.

Guilmet, G. M., 'The Evolution of Tool-Using and Tool-Making Behaviour', *Man* (ns), vol. 12, no. 1, 1977, pp. 38-47.

Hall, K. R. L., 'Tool-Using Performances as Indicators of Behavioral Adaptability', in Jay (ed), 1968, pp. 131-48.

Hallowell, A. I., 'Self, Society and Culture in Phylogenetic Perspective', in Tax (ed), 1960, pp. 309-71.

Hallowell, A. I., 'The Protocultural Foundations of Human Adaptation', in Washburn (ed), 1962, pp. 236-55.

Harding, R. S. O., 'Meat-Eating and Hunting in Baboons', in Tuttle (ed), 1975, pp. 245-57.

Harnad, S. R., Steklis, H. D. and Lancaster, J. (eds), *Origins and Evolution of Language and Speech*, New York Academy of Sciences, vol. 280, 1976.

Hayden, B. (ed), *Lithic Use-Wear Analysis*, New York, Academic Press, 1979.

Hayes, K. J., and Nissen, C. H., 'Higher Mental Functions of a Home Raised Chimpanzee', in Schrier and Stollnitz (eds), 1971, pp. 59-115.

Hewes, G. W., 'Primate Communication and the Gestural Origin of Language', *Current Anthropology*, vol. 14, nos 1-2, 1973, pp. 5-12.

Hewes, G. W., 'Language Origin Theories', in Rumbaugh (ed), 1977, pp. 3-53.

Hockett, C., 'Animal "Languages" and Human Language', in Spuhler (ed), 1959, pp. 32-9.

Hockett, C., 'The Origin of Speech', *Scientific American*, vol. 203, no. 3, 1960, pp. 88-96.

Hockett, C. F., and Ascher, R., 'The Human Revolution', in Montagu (ed), 1968, pp. 20-101.

Hoffman, J., *Marxism and the Theory of Praxis*, London, Lawrence & Wishart, 1975.

Hoffman, J., 'The Dialectics of Nature: "the Natural-Historical Foundation of our Outlook" ', *Marxism Today*, vol. 21, no. 1, 1977, pp. 11-18.

Holloway, R. L., 'Culture: a *Human* Domain', *Current Anthropology*, vol. 10, no. 4, 1969, pp. 395-412.

Holloway, R. L. (ed), *Primate Aggression and Xenophobia*, New York, Academic Press, 1974.

Holloway, R. L., 'The Casts of Fossil Hominid Brains', *Scientific American*, vol. 231, no. 1, 1974, pp. 106-15.

Howell, F. C., and Bourlière, F. (eds), *African Ecology and Human Evolution*, London, Methuen, 1964.

Howells, W., *Evolution of the Genus 'Homo'*, Reading, Mass., Addison-Wesley, 1973.

Ilyenkov, E. V., 'The Concept of the Ideal', in *Philosophy in the USSR: Problems of Dialectical Materialism*, op. cit., 1977, pp. 71-9.

Isaac, G. L., 'The Diet of Early Man: Aspects of Archaeological Evidence from Lower and Middle Pleistocene Sites in Africa', *World Archaeology*, vol. 2, no. 3, 1971, pp. 278-98.

Isaac, G. L., 'Plio-Pleistocene Artifact Assemblages from East Rudolf, Kenya', in Coppens, Howell, Isaac and Leakey (eds), 1976, pp. 552-64.

Isaac, G. L., 'The Food-Sharing Behavior of Protohuman Hominids', *Scientific American*, vol. 238, no. 3, 1978, pp. 90-108.

Isaac, G. L., and McCown, E. R. (eds), *Human Origins*, California, W. A. Benjamin, 1976.

Jay, P. C. (ed), *Primates: Studies in Adaptation and Variability*, New York, Holt, Rinehart & Winston, 1968.

Jerison, H. J., 'Discussion Paper: the Paleoneurology of Language', in Harnad, Steklis and Lancaster (eds), 1976, pp. 370-82.

Johanson, D. C., and Edey, M., *Lucy, the Beginings of Humankind*, London, Granada, 1981.

Johanson, D. C., and Taieb, M., 'Plio-Pleistocene Discoveries in Hadar, Ethiopia', *Nature*, vol. 260, 1976, pp. 293-7.

Johanson, D. C., Taieb, M., Coppens, Y. and Roche, H., 'Expédition internationale de l'Afar, Ethiopie, (4e et 5e campagnes 1975-1977), Nouvelles découvertes d'hominidés et découvertes d'industries lithiques pliocènes à Hadar', *Comptes Rendus de l'Académie des Sciences* (series D.), vol. 287, 1978, pp. 237-40.

Johanson, D. C., and White, T. D., 'A Systematic Assessment of Early African Hominids', *Science*, vol. 203, 1979, pp. 321-30.

Jolly, C. J., 'The Seed-Eaters: a New Model of Hominid Differentiation Based on a Baboon Analogy', *Man* (ns), vol. 5, no. 1, 1970, pp. 5-26.

Jolly, C. J. (ed), *Early Hominids of Africa*, London, Duckworth, 1978.

Jones, P. R., 'Effect of Raw Materials on Biface Manufacture', *Science*, vol. 204, 1979, pp. 935-6.

Jones, P. R., 'Experimental Butchering with Modern Stone Tools and its Relevance for Paleolithic Archaeology', *World Archaeology*, vol. 12, no. 2, 1980, pp. 152-65.

Julkunen, R., 'A Contribution to the Categories of Social Time and the Economy of Time', *Acta Sociologica*, vol. 20, no. 1, 1977, pp. 5-24.

Kawamura, S., 'The Process of Sub-Culture Propagation among Japanese Macaques', in Southwick (ed), 1963, pp. 82-90.

Khrustov, G. F., 'The Problem of the Origin of Man', *Soviet Psychology*, vol. 9, fall, 1970, pp. 6-31.

Kinsey, W. G., 'Evolution of the Human Canine Tooth', *American Anthropologist*, vol. 73, no. 3, 1971, pp. 680-94.

Klejn, L.S., 'A Panorama of Theoretical Archaeology', *Current Anthropology*, vol. 18, no. 1, 1977, pp. 1-42, vol. 18, no. 2, 1977, pp. 371-3.
Kochetkova, V.I., *Paleoneurology* (trans. and ed by H.J. and I. Jerison), New York, John Wiley, 1978.
Kolb, B., and Whishaw, I.Q., *Fundamentals of Human Neuropsychology*, San Francisco, W.H. Freeman, 1980.
Kortlandt, A., discussion contribution on 'Meat-Eating and Behavioral adaptations to Hunting', in Tuttle (ed), 1975, p. 302.
Kretzoi, M., 'New Ramapithecines and *Pliopithecus* from the Lower Pliocene of Rudâbanya in North Eastern Hungary', *Nature*, vol. 257, 1975, pp. 578-81.
Lancaster, J.B., 'On the Evolution of Tool-Using Behavior', *American Anthropologist*, vol. 70, no. 1, 1968, pp. 56-66.
Lancaster, J.B., 'Primate Communication Systems and the Emergence of Human Language', in Jay (ed), 1968, pp. 439-57.
Lancaster, J.B., *Primate Behavior and the Emergence of Human Culture*, New York, Holt, Rinehart & Winston, 1975.
Laughlin, W.S., 'Hunting: an Integrating Biobehavioral System and its Evolutionary Importance', in Lee and De Vore (eds), 1968, pp. 304-20.
Leacock, E.B., editor's introduction to F. Engels, 'The Part Played by Labor in the Transition from Ape to Man', in Engels, op. cit., 1972, pp. 245-9.
Leakey, L.S.B., 'A New Fossil Skull from Olduvai', *Nature*, vol. 184, 1959, pp. 491-3.
Leakey, L.S.B., 'Finding the Earliest Man', *National Geographic*, vol. 118, September, 1960, pp. 420-35.
Leakey, L.S.B., 'Adventures in the Search for Man', *National Geographic*, vol. 123, January, 1963, pp. 132-52.
Leakey, M., *Olduvai Gorge*, vol. 3, Cambridge University Press, 1971.
Leakey, M.D., and Hay, R.L., 'Pliocene Footprints in the Laetolil Beds at Laetoli, Northern Tanzania', *Nature*, vol. 278, 1979, pp. 317-23.
Leakey, R.E.F., 'Evidence for an Advanced Plio-Pleistocene Hominid from East Rudolf, Kenya', *Nature*, vol. 242, 1973, pp. 447-50.
Leakey, R.E.F., 'Skull 1470', *National Geographic*, vol. 143, June, 1973, pp. 818-29.
Leakey, R.E.F., 'The Making of Mankind', the *Listener*, vol. 105, no. 2711, 7 May 1981, pp. 598-601.
Leakey, R.E., and Lewin, R., *Origins*, London, Macdonald & Jane's, 1977.
Leakey, R., and Lewin, R., *People of the Lake*, London, Collins, 1979.
Lee, R.B., 'What Hunters do for a Living, or How to Make Out on Scarce Resources', in Lee and De Vore (eds), 1968, pp. 30-48.
Lee, R.B., *The !Kung San*, Cambridge University Press, 1979.
Lee, R.B., 'Appendix B. Plants of the Dobe area', in Lee, op. cit., 1979, pp. 464-73.
Lee, R.B., 'Appendix E. The Hand-to-Mouth Existence: a Note on the Origin of the Human Economy', in Lee, op. cit., 1979, pp. 489-94.
Lee, R.B., and De Vore, I. (eds), *Man the Hunter*, Chicago, Aldine, 1968.
Lenin, V.I., *Materialism and Empirio-criticism*, Collected Works, vol. 14,

London, Lawrence & Wishart, 1962.
Lenin, V.I., *Philosophical Notebooks,* Collected Works, vol. 38, London, Lawrence & Wishart, 1962.
Leontiev, A.A., 'Social and Natural Semiotics', in Morton (ed), 1971, pp. 122-30.
Leontiev, A.A., 'Sense as a Psychological Concept', in Prucha (ed), 1976, pp. 82-90.
Leontiev, A.A., 'Some New Trends in Soviet Psycholinguistics', *Soviet Psychology,* vol. 25, no. 2, winter, 1976-7, pp. 15-25.
Leontiev, A.N., 'The Problem of Activity in Psychology', *Soviet Psychology,* vol. 13, no. 2, winter, 1974-5, pp. 4-33.
Leontiev, A.N., *Activity, Consciousness and Personality,* Englewood Cliffs, N.J., Prentice-Hall, 1978.
Leontiev, A.N., and Leontiev, A.A., 'The Social and the Individual in Language', *Language and Speech,* vol. 2, no. 4, 1959, pp. 193-204.
Leontiev, A.N., and Luria, A.R., 'The Psychological Ideas of L.S. Vygotskii', in Wolman (ed), 1968, pp. 338-67.
Levine, N., *The Tragic Deception: Marx contra Engels,* Oxford, Clio Books, 1975.
Lewin, R., 'Evolutionary Theory under Fire', *Science,* vol. 201, 1980, pp. 883-7.
Lewis, G.E., 'Preliminary Notice of New Man-Like Apes from India', *American Journal of Science* (5th series), vol. 27, no. 159, 1934, pp. 161-82.
Lewis, J., *The Uniqueness of Man,* London, Lawrence & Wishart, 1974.
Lieberman, P., Crelin, E.S. and Klatt, B.M., 'Phonetic Ability and Related Anatomy of the Newborn and Adult Human, Neanderthal Man, and the Chimpanzee', *American Anthropologist,* vol. 74, no. 3, 1972, pp. 287-307.
Limber, J., 'Language in Child and Chimp?', in Sebeok and Umiker-Sebeok (eds), 1980, pp. 197-220.
Lomov, B.F., 'Lenin's Theory of Reflection and Psychology', *Studia Psychologica,* vol. 13, no. 3, 1971, pp. 173-9.
Lovejoy, C.O., Burstein, A.H. and Heiple, K.G., 'Primate Phylogeny and Immunological Distance', *Science,* vol. 176, 1972, pp. 803-5.
Lovejoy, C.O., Heiple, G.H. and Burstein, A.H., 'The Gait of *Australopithecus*', *American Journal of Physical Anthropology,* vol. 38, no. 3, 1973, pp. 757-80.
Loy, J., 'The Descent of Dominance in *Macaca:* Insights into the Structure of Human Societies', in Tuttle (ed), 1975, pp. 153-80.
Lukacs, G., *History and Class Consciousness,* London, Merlin Press, 1971.
Luria, A.R., 'The Role of Language in the Formation of Temporary Connections', in Simon (ed), 1957, pp. 115-29.
Luria, A.R., *The Role of Speech in the Regulation of Normal and Abnormal Behaviour,* Oxford, Pergamon Press, 1961.
Luria, A.R., *The Working Brain,* Hamondsworth, Penguin, 1973.
Luria, A.R., 'Scientific Perspectives and Philosophical Dead Ends in Modern Linguistics', *Cognition,* vol. 3, no. 4, 1974-5, pp. 377-85.
McGrew, W.C., and Tutin, C.E.G., 'Evidence for a Social Custom in Wild Chimpanzees?', *Man* (ns), vol. 13, no. 2, 1978, pp. 234-51.

McGrew, W.C., Tutin, C.E.G. and Baldwin, P.J., 'Chimpanzees, Tools and Termites: Cross Cultural Comparisons of Senegal, Tanzania and Rio Muni', *Man* (ns), vol. 14, no. 2, 1979, pp. 185-214.
McLeish, J., *Soviet Psychology: History, Theory, Content*, London, Methuen, 1975.
Mann, A., 'Hominid and Cultural Origins', *Man* (ns), vol. 7, no. 3, 1972, pp. 379-96.
Marler, P., 'Communication in Monkeys and Apes', in De Vore (ed), 1965, pp. 544-84.
Marler, P., and Tenaza, R., 'Signalling Behavior of Apes with Special Reference to Vocalization', in Sebeok (ed), 1977, pp. 965-1003.
Marx, K., *Economic and Philosophical Manuscripts of 1844*, London, Lawrence & Wishart, 1970.
Marx, K., *The German Ideology*, London, Lawrence & Wishart, 1970.
Marx, K., *Capital*, vol. 1, London, Lawrence & Wishart, 1972.
Marx, K., *Grundrisse*, Hammondsworth, Penguin, 1973.
Marx, K., and Engels, F., *Selected Correspondence*, Moscow, Progress Publishers, 1975.
Marx, K., letter to Ludwig Kugelmann in Hanover, 11 July 1868, in Marx and Engels, op. cit., 1975, pp. 195-7.
Menzel, E.W. (ed), *Precultural Primate Behavior, IVth International Congress of Primatology*, vol. 1, Basel, Krager, 1973.
Montagu, M.F.A. (ed), *Culture: Man's Adaptive Dimension*, Oxford University Press, 1968.
Montagu, M.F.A., 'Toolmaking, Hunting and the Origin of Language', in Harnad, Steklis and Lancaster (eds), 1976, pp. 266-74.
Montagu, M.F.A. (ed), *Learning Non-aggression: The Experience of Pre-literate Societies*, Oxford University Press, 1978.
Montagu, M.F.A. (ed), *Sociobiology Examined*, Oxford University Press, 1980.
Morton, J. (ed), *Biological and Social Factors in Psycholinguistics*, London, Logos Press, 1971.
Mounin, G., 'Language, Communication, Chimpanzees', *Current Anthropology*, vol. 17, no. 1, 1976, pp. 1-21.
Napier, J., 'The Evolution of the Hand', *Scientific American*, vol. 207, no. 6, 1962, pp. 56-62.
Napier, J., 'The Antiquity of Human Walking', *Scientific American*, vol. 216, no. 4, 1967, pp. 56-66.
Nishida, T., review of B.B. Beck, 1980, *Nature*, vol. 289, 1981, p. 616.
Noiré, L., *The Origin and Philosophy of Language*, Chicago, The Open Court Publishing Company, 1917.
Oakley, K.P., 'Tools Makyth Man', *Antiquity*, vol. 31, no. 124, 1957, pp. 199-209.
Oakley, K.P., 'On Man's Use of Fire, with Comments on Tool-Making and Hunting', in Washburn (ed), 1962, pp. 176-93.
Oakley, K.P., 'Man the Skilled Toolmaker', *Antiquity*, vol. 43, no. 171, 1969, pp. 222-4.
Oppenheimer, J.R., 'Communication in New World Monkeys', in Sebeok (ed), 1977, pp. 851-89.

Patsch, G., *Grundfragen Der Sprachtheorie,* Halle, Velag Max Niemeyer, 1955.
Pavlov, I. P., *Selected Works,* Moscow, Foreign Languages Publishing House, 1957.
Penfield, W., and Rasmussen, T., *The Cerebral Cortex of Man,* New York, Macmillan, 1950.
Peters, R., and Mech, D. L., 'Behavioral and Intellectual Adaptations of Selected Mammalian Predators to the Problem of Hunting Large Animals', in Tuttle (ed), 1975, pp. 279-300.
Philbeam, D., Meyer, G. F., Badgley, C., Rose, M. D., Pickford, M. K. L., Behrensmeyer, A. K. and Shah, S. M. I., 'New Hominoid Primates from Siwaliks of Pakistan and their Bearing on Hominoid Evolution', *Nature,* vol. 270, 1977, pp. 689-95.
Philosophy in the USSR: Problems of Dialectical Materialism (collective author), Moscow, Progress Publishers, 1977.
Premack, D., 'Language in Chimpanzee?', *Science,* vol. 172, 1971, pp. 808-22.
Premack, D., 'On the Assessment of Language Competence in the Chimpanzee', in Schrier and Stollnitz (eds), 1971, pp. 185-228.
Premack, D., *Intelligence in Ape and Man,* Hillsdale, New Jersey, Lawrence Erlbaum Associates, 1976.
Prucha, J. (ed), *Soviet Studies in Language and Language Behaviour,* Amsterdam and Oxford, North-Holland Publication Co., 1976.
Rahmani, L., *Soviet Psychology,* New York, International Universities' Press, 1976.
Reader, J., *Missing Links,* London, Collins, 1981.
Read-Martin, C. E., and Read, D. W., 'Australopithecine Scavenging and Human Evolution: an Approach from Faunal Analysis' *Current Anthropology,* vol. 16, no. 3, 1975, pp. 359-68.
Reed, E., *Woman's Evolution,* New York, Pathfinder Press, 1975.
Reed, E., *Sexism & Science,* New York, Pathfinder Press, 1978.
Reed, E., 'Evolutionism and Anti-Evolutionism' (1957), in Reed, op. cit., 1978, pp. 161-77.
Reed, E., 'Primates and Prejudice', in Reed, op. cit., 1978, pp. 8-33.
Reiter, R. R. (ed), *Toward an Anthropology of Women,* New York, Monthly Review Press, 1975.
Révész, G., *The Origins and Prehistory of Language,* London, Longman's Green & Company, 1956.
Rieber, R. W. (ed), *The Neuropsychology of Language: Essays in Honor of Eric Lenneberg,* New York, Plenum Press, 1976.
Riegel, K. F., and Rosenwald, G. C. (eds), *Structure and Transformation: Developmental and Historical Aspects,* New York, John Wiley & Sons, 1975.
Riegel, K. F., 'Structure and Transformation in Modern Intellectual History', in Riegel and Rosenwald (eds), 1975, pp. 3-45.
Robinson, B. W., 'Vocalization Evoked from Forebrain in *Macaca Mulatta*', *Physiology and Behavior,* vol. 2, no. 4, 1967, pp. 345-54.
Roe, A., 'Psychological Definitions of Man', in Washburn (ed), 1963, pp. 320-31.
Rumbaugh, D. M. (ed), *Language Learning by a Chimpanzee: The Lana Project,*

New York, Academic Press, 1977.

Rumbaugh, D. M., Van Glasersfeld, E. C., Gill, T. V., Warner, H., Pisani, P., Brown, J. V. and Bell, C. L., 'The Language Skills of a Young Chimpanzee in a Computer Controlled Training Situation', in Tuttle (ed), 1975, pp. 391-401.

Ruse, M., *Sociobiology: Sense or Nonsense?*, Dordrecht, Holland, Riedel, 1979.

Sacks, K., *Sisters and Wives*, Westport, Conn., Greenwood Press, 1979.

Sahlins, M., *Stone Age Economics*, London, Tavistock, 1974.

Sahlins, M., *The Use and Abuse of Biology*, London, Tavistock, 1977.

Sarich, U. M., and Wilson, A. C., 'Immunological Time Scale for Hominid Evolution', *Science*, vol. 158, 1967, pp. 1200-3.

Savage-Rumbaugh, E. S., Rumbaugh, D. M. and Boysen, S., 'Linguistically Mediated Tool Use and Exchange by Chimpanzees (Pan Troglodytes)', *The Behavioral and Brain Science*, vol. 4, 1978, pp. 539-54; reprinted in Sebeok and Umiker-Sebeok (eds), 1980, pp. 353-83.

Schrier, A. M., and Stollnitz, F. (eds), *Behavior of Nonhuman Primates: Modern Research Trends*, vol. 4, New York, Academic Press, 1971.

Science as Ideology Group of the British Society for Social Responsibility in Science, 'The New Synthesis is an Old Story', *New Scientist*, vol. 70, no. 1000, 1976, pp. 346-8.

Scientific American, The Brain, articles republished from *Scientific American*, vol. 241, no. 9, 1979, San Francisco, W. H. Freeman, 1979.

Sebeok, T. A. (ed), *How Animals Communicate*, Indiana University Press, 1977.

Sebeok, T. A., and Umiker-Sebeok, J. (eds), *Speaking of Apes: A Critical Anthology of Two-Way Communication with Man*, New York, Plenum Press, 1980.

Semenov, S. A., *Prehistoric Technology*, London, Cory, Adam & Mackay, 1964.

Semenov, Yu. I., 'The Doctrine of Morgan, Marxism and Contemporary Ethnography', *Soviet Anthropology and Archaeology*, vol. 4, no. 2, fall, 1965, pp. 3-15.

Simon, B. (ed), *Psychology in the Soviet Union*, London, Routledge & Kegan Paul, 1957.

Simons, E. L., 'The Early Relatives of Man', *Scientific American*, vol. 211, no. 1, 1964, pp. 50-62.

Simons, E. L., *'Ramapithecus', Scientific American*, vol. 236, no. 5, 1977, pp. 28-35.

Slocum, S., 'Woman the Gatherer', in Reiter (ed), 1975, pp. 36-50.

Sociobiology Study Group of Science for the People, 'Sociobiology: Another Biological Determinism', *BioScience*, vol. 26, no. 3, 1976, pp. 182-6; reprinted in Caplan (ed), 1978, pp. 280-90.

Southwick, C. H. (ed), *Primate Social Behavior*, Princeton, N.J., Van Nostrand, 1963.

Spirkin, A., 'Consciousness', in Blakeley (ed), 1975, pp. 59-73.

Spuhler, J. N. (ed), 'The Evolution of Man's Capacity for Culture', *Human Biology*, vol. 31, no. 1, 1959.

Stiles, D. N, Hay, R. L. and O'Neill, J. R., 'The MNK Chert Factory Site,

Olduvai George, Tanzania', *World Archaeology*, vol. 5, no. 3, 1974, pp. 285-308.
Struhsaker, T. T., 'Auditory Communication among Vervet Monkeys (Cercopithecus Aethiops)', in Altmann (ed), 1967, pp. 281-324.
Susman, R. L., and Creel, N., 'Functional and Morphological Affinities of the Subadult Hand (OH7) from Olduvai Gorge', *American Journal of Physical Anthropology*, vol. 51, no. 3, 1979, pp. 311-32.
Suzuki, A., 'The Origin of Hominid Hunting: a Primatological Perspective', in Tuttle (ed), 1975, pp. 259-78.
Szalay, F. S., 'Hunting–Scavenging Protohominids: a Model for Hominid Origins', *Man* (ns), vol. 10, no. 3, 1975, pp. 420-9.
Tattersall, J., and Eldredge, N., 'Fact, Theory and Fantasy in Human Paleontology', *American Scientist*, vol. 65, no. 2, 1972, pp. 204-11.
Tax, S. (ed), *Evolution after Darwin*, vol. 2, University of Chicago Press, 1960.
Tax, S. (ed), *Horizons of Anthropology*, London, Allen & Unwin, 1965.
Teleki, G., *The Predatory Behavior of Wild Chimpanzees*, Bucknell University Press, 1973.
Teleki, G., 'The Omnivorous Chimpanzee' *Scientific American*, vol. 288, no. 1, 1973, pp. 32-47.
Teleki, G., 'Chimpanzee Subsistence Technology: Materials and Skills', *Journal of Human Evolution*, vol. 3, 1974, pp. 575-94.
Terrace, H. S., *Nim*, London, Eyre Methuen, 1980.
Terrace, H. S., 'Is Problem-Solving Language?' in Sebeok and Umiker-Sebeok (eds), 1980, pp. 385-405.
Terrace, H. S., Pettito, L. A., Saunders, R. J. and Bever, T. G., 'Can an Ape Create a Sentence?' *Science*, vol. 206, 1979, pp. 891-906.
Thompson, C. R., and Church, R. M., 'An Explanation of the Language of a Chimpanzee', *Science*, vol. 208, 1980, pp. 313-14.
Thomson, G., *The Human Essence*, London, China Policy Study Group, 1974.
Tikhomirov, O. K., review of B. F. Skinner, *Verbal Behavior*, *Word*, vol. 15, 1959, pp. 362-7.
Tobias, P. V., '*Australopithecus*, *Homo Habilis*, Tool-Using and Tool-Making', *South African Archaeological Bulletin*, vol. 20, December. 1965, pp. 167-92.
Toulmin, S., 'The Mozart of Psychology', *New York Review of Books*, vol. 25, no. 14, September, 1978, pp. 51-7.
Trigger, B. G., 'Engels on the Part Played by Labor in the Transition from Ape to Man: an Anticipation of Contemporary Anthropological Theory', *Canadian Review of Sociology and Anthropology*, vol. 4, no. 3, 1966, pp. 165-76.
Turnbull, C., contribution to discussion on 'Primate Behavior and the Evolution of Aggression', in Lee and De Vore (eds), 1968, p. 341.
Tuttle, R. H. (ed), *Socioecology and Psychology of Primates*, The Hague, Mouton, 1975.
Vygotsky, L. S., *Thought and Language* (trans. and ed by E. Hanfmann and G. Vakar), Harvard, M I T Press, 1962.
Vygotsky Memorial Issue, *Soviet Psychology*, vol. 5, no. 3, fall, 1967.
Vygotsky, L. S., 'The Development of Higher Psychological Functions', *The Soviet Review*, vol. 18, no. 3, fall, 1977, pp. 38-51.

Vygotsky, L. S., *Mind in Society: The Development of Higher Psychological Processes* (trans. and ed by M. Cole, V. John-Steiner, S. Scribner and E. Souberman), Harvard University Press, 1978.

Vygotsky, L. S., 'Internalization of Higher Psychological Functions', in Vygotsky, op. cit., 1978, pp. 52-7.

Vygotsky, L. S., 'Tool and Symbol in Child Development', in Vygotsky, op. cit., 1978, pp. 19-30.

Walker, A., and Leakey, R. E. F., 'The Hominids of East Turkana', *Scientific American*, vol. 239, no. 2, 1978, pp. 44-56.

Washburn, S. L., 'Speculations on the Interrelations of the History of Tools and Biological Evolution', in Spuhler (ed), 1959, pp. 21-31.

Washburn, S. L., 'Tools and Human Evolution', *Scientific American*, vol. 203, no. 3, 1960, pp. 63-73.

Washburn, S. L. (ed), *Social Life of Early Man*, London, Methuen, 1962.

Washburn, S. L. (ed), *Classification and Human Evolution*, Chicago University Press, 1963.

Washburn, S. L., discussion contribution on 'Primate Behavior and the Evolution of Aggression', in Lee and De Vore (eds), 1968, p. 342.

Washburn, S. L., 'Human Behavior and the Behavior of Other Animals', *American Psychologist*, vol. 33, May, 1978, pp. 405-18.

Washburn, S. L., and Benedict, B., 'Nonhuman Primate Culture', *Man* (ns), vol. 14, no. 1, 1979, pp. 163-4.

Washburn, S. L., and De Vore, I., 'Social Behavior of Baboons and Early Man', in Washburn (ed), 1962, pp. 91-105.

Washburn, S. L., and Lancaster, C. S., 'The Evolution of Hunting', in Lee and De Vore (eds), 1968, pp. 293-303.

Washburn, S. L., and Lancaster, J. B., 'On Evolution and the Origin of Language', *Current Anthropology*, vol. 12, no. 3, 1971, pp. 384-6.

Weitz, S. (ed), *Nonverbal Communication*, Oxford University Press, 1974.

Wells, H. K., *Ivan P. Pavlov: Toward a Scientific Psychology and Psychiatry*, London, Lawrence & Wishart, 1956.

White, L. A., 'On the Use of Tools by Primates', *Journal of Comparative Psychology*, vol. 34, no. 3, 1942, pp. 369-74.

White, L. A., *The Science of Culture*, New York, Grove Press, 1949.

White, L. A., *The Evolution of Culture*, New York, McGraw-Hill, 1959.

Wind, J., 'Phylogeny of the Human Vocal Tract', in Harnad, Steklis and Lancaster (eds), 1976, pp. 612-30.

Wing, E. J., and Brown, A. B. (eds), *Paleonutrition*, New York, Academic Press, 1980.

Wilson, E. O., *Sociobiology: The New Synthesis*, Harvard University Press, 1975.

Wilson, E. O., *On Human Nature*, Harvard University Press, 1978.

Wolberg, D., 'The Hypothesised *Osteodontokeratic* Culture of the Australopithecinae: a Look at the Evidence and the Opinions', *Current Anthropology*, vol. 11, no. 1970, pp. 23-30.

Wolfgang, A. (ed), *Nonverbal Behavior: Applications and Cultural Implications*, New York, Academic Press, 1979.

Wolman, B. (ed), *Historical Roots of Contemporary Psychology*, New York, Harper & Row, 1968.

Wolpoff, M. H., 'Analogies and Interpretation in Palaeoanthropology', in Jolly (ed), 1978, pp. 461-502.
Woodburn, J., 'Stability and Flexibility in Hadza Residential Groups', in Lee and De Vore, (eds), 1968, pp. 103-10.
Woolfson, C., 'Culture, Language and the Human Personality', *Marxism Today,* vol. 21, no. 8, 1977, pp. 229-40.
Wright, R. V. S., 'Imitative Learning of a Flake Stone Technology – the Case of an Orangutang', *Mankind,* vol. 8, no. 4, 1972, pp. 296-304.
Zavadskii, K. M., Georgievskii, A. M. and Mozelov, A. P., 'Engels and Darwinism', *Soviet Studies in Philosophy,* vol. 10, no. 1, summer, 1971, pp. 63-80.

Name index

Subject index

For Product Safety Concerns and Information please contact our EU representative GPSR@taylorandfrancis.com Taylor & Francis Verlag GmbH, Kaufingerstraße 24, 80331 München, Germany

Batch number: 10725798

Printed by Printforce, the Netherlands